M000192133

INTRODUCTION TO STORMWATER

INTRODUCTION TO STORMWATER

Concept, Purpose, Design

BRUCE K. FERGUSON

JOHN WILEY & SONS, INC.

New York / Chichester / Weinheim / Brisbane / Singapore / Toronto

This text is printed on acid-free paper.

Copyright © 1998 by John Wiley & Sons, Inc.

All rights reserved. Published simultaneously in Canada.

No part of this publication may be reproduced, stored in a retrieval system or transmitted in any form or by any means, electronic, mechanical, photocopying, recording, scanning or otherwise, except as permitted under Sections 107 or 108 of the 1976 United States Copyright Act, without either the prior written permission of the Publisher, or authorization through payment of the appropriate per-copy fee to the Copyright Clearance Center, 222 Rosewood Drive, Danvers, MA 01923, (508) 750-8400, fax (508) 750-4744. Requests to the Publisher for permission should be addressed to the Permissions Department, John Wiley & Sons, Inc., 605 Third Avenue, New York, NY 10158-0012, (212) 850-6011, fax (212) 850-6008, E-Mail: PERMREQ @ WILEY.COM.

This publication is designed to provide accurate and authoritative information in regard to the subject matter covered. It is sold with the understanding that the publisher is not engaged in rendering legal, accounting, or other professional services. If legal advice or other expert assistance is required, the services of a competent professional person should be sought.

Library of Congress Cataloging-in-Publication Data

Ferguson, Bruce K.
   Introduction to stormwater : concept, purpose, design /
  Bruce Ferguson.
      p.   cm.
   Includes index.
   ISBN 0-471-16528-X (cloth : alk. paper)
   1. Urban runoff—Management.   2. City planning—Environmental
  aspects. I. Title.
  TD657.F47   1998
  628'.212—dc21   97-20304

Printed in the United States of America

10 9 8 7 6 5 4 3 2 1

CONTENTS

PREFACE

Stormwater management lies near the heart of basic landscape architecture and site engineering. Professional ethics demand that every practitioner integrate stormwater safely and meaningfully with every urban community and ecosystem. To do it well is to preserve both the functioning of urban watersheds and the quality of life of the people who live within them.

Stormwater is also a special area of potentially deep knowledge and extensive skill. It changes as scientific knowledge expands. A practitioner can spend an entire career focusing on this area and still be discovering new relationships and new skills after decades of practice and research.

Essentially all site developments, of all kinds, involve impervious and compacted surfaces. The change in land cover increases runoff over the surface, dumps flood waters into streams, reduces groundwater recharge, diverts water from base flows, and turns oil from the streets into pollutants.

But water quality must be protected, floods suppressed, droughts prevented, water conserved. And the infrastructure that accomplishes these goals must be safe and resourceful. These demands are reflected in specific stormwater and development laws. More universally, the demands are reflected in courts of law, where people whose health and safety are compromised have a right to be compensated.

This book is therefore aimed at people who have a job to do in a world characterized by deepening environmental issues and developing technologies. It gives you both the "why" and the "how" of stormwater. It supplements broad basic training in grading, drainage, site construction, natural science, and site design with fundamental concepts of stormwater hydrology and their implications for urban design and the environment. The presentation is graphic and intuitive. At the same time, sufficient mathematics are presented to allow the user to perceive the underlying theory and to implement the quantitative procedures in spreadsheets, geographic information systems (GIS), or other software. This is a textbook for university classes that emphasize the integration of the science of hydrology with the art of design, as well as a reference for practitioners who need to review and update their applied skills.

The "why" is addressed in the first two chapters, which describe the role of stormwater in the natural environment and in human experience. The next three chapters give the background of quantitative hydrologic concepts and the analysis of hydrologic flows that constrain design. The "how" is discussed in the last five chapters, which introduce the functional design options for stormwater, one option per chapter. A complete range of management options are described in an easy-to-learn manner, taking the reader rapidly to the completion of applied exercises.

The purpose of this book is to give its users an understanding of the relationships of stormwater to the human and natural environment, recognition of the range of available management approaches and their implications for water resources and site development, and skill in applying basic quantitative methods to estimate and design for stormwater.

This book prepares practitioners specifically:

1. To prepare complete practical applications for simple projects such as small on-site facilities. The charts, data, and methods in this book are adequate in many cases, although details of data and procedures must be replaced with local practices where required. The basic skills acquired in completing the book's exercises are sufficient to prepare users to move to other stormwater methods and data where required in practice.
2. To understand, anticipate, and deal with stormwater management objectives and constraints in the master planning of large projects. The text and exercises in this book introduce the arguments for and against different management options in varying site conditions. A site planner who understands stormwater objectives and constraints can communicate constructively with the specialists who carry out the hydraulic details. Only such a person can integrate these objectives and constraints supportively into a functioning master plan.

Many chapters include brief analytical exercises that penetrate the logic underlying important concepts, as well as a significant computational exercise for each type of hydrologic procedure that is discussed. Computational exercises later in the book use the results of earlier ones. Solving each exercise for two sites that contrast in soil or slope stimulates discussion and understanding of the effects of these factors on hydrology and design. A class can divide into two groups, each group completing the exercise for one site. You can use sites that you are working on in practice, typical sites within your region with which you are familiar, or the hypothetical sites described in the Appendix.

The references listed at the end of each chapter include the sources of data directly cited in the text, plus some selected additional references that are primary classics in the field or that can benefit the users of this book by providing further elucidation of the points made here.

This book is not an exhaustively detailed stormwater reference. The level of quantitative hydrology that it covers is basic. There are plenty of detailed hydrologic references available; some are listed at the end of this preface. Further sophistication in hydrologic theory or modeling requires specialized professional courses, resident graduate training, or extensive specialized experience.

This book does not cover the physical materials of stormwater structures such as dams and weirs. Carrying out a stormwater program through site layout and construction remains a matter for broad design competence, beyond the specialized field of hydrology.

The responsibility for applying any method to a specific site rests with the professional person in charge of the project. Regulations and accepted practices vary from region to region, and specific needs vary from site to site. Locally valid standards and practices should be used for site-specific design.

The quantitative calculations that are taught in this book are useful and necessary tools. But it is important not to be distracted by technicalities. I want you to understand fundamentally what you are doing and why. The point is not a number. The point is what you are doing to the land. The goal is to solve human and environmental problems, using whatever kinds of tools are required at each stage of conception, implementation, and evaluation.

In stormwater management, new approaches have been born, older approaches have been reevaluated, and the "basics" are not what they used to be. It is the responsibility of every practitioner to maintain the state of the practice at the state of the science. I offer this book as a reflection of the reciprocal evolutions of the science of hydrology and the art of design.

In doing so, I owe an intellectual debt to Ian McHarg, Narendra Juneja, Luna Leopold, Peter McCleary, Hollister Kent, Robert Giegengack, Seymour Subitsky, Michael Dufalla, and Ruth Patrick, and an ethical and motivational debt, as always, to Albert B. Ferguson, Jr.

REFERENCES

Debo, Thomas N., and Andrew Reese, 1975, *Municipal Stormwater Management*, Boca Raton: Lewis.

Dunne, Thomas, and Luna B. Leopold, 1978, *Water in Environmental Planning*, San Francisco: Freeman.

Field, Richard, Marie L. O'Shea, and Kee Kean Chin, 1993, *Integrated Stormwater Management*, Boca Raton: Lewis.

Malcom, H. Rooney, 1989, *Elements of Urban Stormwater Design*, Raleigh: North Carolina State University Industrial Extension Service.

McCuen, Richard H., 1989, *Hydrologic Analysis and Design*, Englewood Cliffs, N.J.: Prentice-Hall.

Schueler, Thomas R., 1987, *Controlling Urban Runoff: A Practical Manual for Planning and Designing Urban BMPs*, Washington: Metropolitan Washington Council of Governments.

Stahre, Peter, and Ben Urbonas, 1990, *Stormwater Detention for Drainage, Water Quality, and CSO Management*, Englewood Cliffs, N.J.: Prentice Hall.

Tourbier, J. Toby, and Richard Westmacott, 1981, *Water Resources Protection Technology, A Handbook of Measures to Protect Water Resources in Land Development*, Washington: Urban Land Institute.

CHAPTER 1

STORMWATER AND ENVIRONMENT

Stormwater is not a mechanical system. It is an environmental process, joining the atmosphere, the soil, vegetation, land use, and streams, and sustaining landscapes. In every landscape the falling of the rain, the shining of the sun and the blowing of the wind are the beginning of all life.

ENVIRONMENTAL PROCESS

When left undisturbed, soil and vegetation evolve to absorb rain and make it part of the living ecosystem. Since before the first human beings walked on the earth on two feet, nature has been infiltrating water. Roots of grasses and trees reach into the soil; root hairs separate particles of clay; ants and beetles excavate voids in the soil mass; roots decompose, leaving networks of macropores; leaves fall from the trees each year to form a mulch over the soil; earthworms pull the leaves into their burrows, where they ingest them and add their matter to the soil structure; the boles fall to the earth and feed mosses as they decompose. Mineral soil is made open and porous; clay takes on the permeability of gravel. This is a system that accepts and absorbs rain. Nature is working to restore this kind of soil wherever natural processes are given a chance to work freely.

By accepting and absorbing rainfall, the native environment maintains its equilibrium and its health. Organic matter and soil pores suspend the water in the soil, making it available to the roots of native plants. They filter out passing solid particles and build them into the soil matrix. Storage in the soil turns intermittent pulses of rainfall into a perennial moisture supply. Microorganisms decompose pollutants and turn them into nutrients for the living system. Deeper below, sheets and pools of groundwater discharge to streams slowly, almost steadily, months after the rain falls, to the streams and wetlands where aquatic organisms survive over dry summers. Over all the area of every watershed, in the pores of

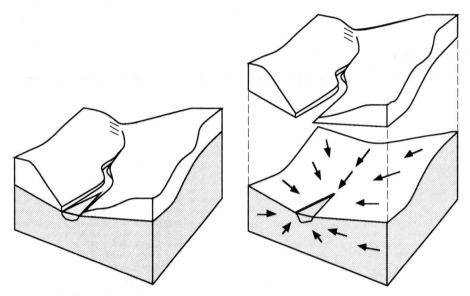

Figure 1.1 The storage and discharge of groundwater in a watershed.

the soil, there is more storage than the Corps of Engineers could create by damming the rivers (Figure 1.1).

In streams, base flow reflects the gradual discharge of water that soaked into the subsurface and was held in the extensive periods between storms. It can be seen in the low to moderate flow of seasonal and perennial streams weeks after storms have passed. In rainy periods, hydraulic gradient increases and subsurface discharge accelerates, releasing water from storage. In dry months, hydraulic gradient declines, conserving in the landscape the water that remains.

In contrast, direct runoff originates on impervious surfaces and saturated soils. Direct runoff can be seen in the rise of streams during storms and in the sheets and rivulets of surface runoff all over developed urban sites far above the springs that supply base flow to seasonal and perennial streams. This is the immediate result of short-term rainfall; it can reach a stream in minutes.

Figure 1.2 shows an example of how base flows and storm flows combine in a stream. The figure is a hydrograph, with time given on the horizontal axis and rate of flow on the vertical axis. The curve of daily discharge in Atlanta's Peachtree Creek shows a pattern of long, low, more or less steady base flows, rising and falling gently with the seasons, punctuated by short, intense bursts of direct runoff accompanying storms. After each storm, the flow falls and merges gradually into the general base level as subsurface processes take over from surface processes. The total volume of water discharging over the course of a year is a composite of base flows and storm flows.

The channels and floodplains that carry base flows and storm flows are dynamic landscapes, where interacting biotic and physical systems are constantly regrowing and readjusting. Riparian wetlands can be ecologically the most productive of all natural communities in a region. In undisturbed streams, flood waves rise frequently but gently over their banks. Flood waters stay long on the floodplains, where they attenuate and infiltrate. From the alluvial aquifer, groundwater discharges gradually, sustaining base flows. Mature stream systems tend to maintain approximate equilibrium during small perturbations and

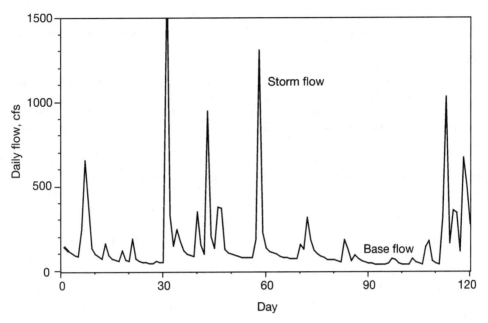

Figure 1.2 Daily discharge in Peachtree Creek, Atlanta, Georgia, March through June 1960 (from U.S. Geological Survey gauge record).

gradual long-term change. When watershed or riparian circumstances change, they adjust by aggrading, degrading, and meandering. A large flood temporarily damages habitat by crushing organisms with tumbling rocks, stranding them on floodplains, and flushing out stored biomass. But floods also renew habitat by digging pools where fish rest and flushing silt from the interstices of rocks where insects seek cover.

To people, it is of great benefit—or at least potential benefit—that nature evolves to work that way. Some persons refer to the work of the earth as "environmental services" and assign dollar values them. Water that infiltrates the soil supports our plants, whether crops, maintained urban landscapes, or native prairies and forests. Where we allow the system to work, water that infiltrates the soil replenishes the aquifers from which we take our well water. Its gradual discharge from the earth makes floods moderate. The streams and reservoirs are full of water and fish, the wetlands are sustained, and erosion is unknown. These naturally sustained qualities decrease care, work, and worry; where they are allowed to persist, they make the need for controls and replacement systems exceptional rather than routine.

Urban development has been changing all that. Impervious pavements are collection pans that concentrate runoff and all the pollutants that accumulate on them, and propel everything immediately into streams without treatment.

WHERE WE'VE COME FROM

A hundred years ago, of course, there were fewer people in the United States and our cities were smaller. But the forms we gave our cities and the ways we lived in them helped keep us out of nature's way.

At that time, railroads and streetcars moved large numbers of people cheaply and cleanly within and between towns. Walking was the other major means of transportation, and urban land use combinations evolved within the constraints of daily walking distances. Development in the suburbs concentrated within walking distance of trolley lines.

City streets were paved with cobbles and bricks, permeable to small amounts of rainfall and runoff. Minor residential streets had no curbs; they were flanked by swales or ditches that kept streets drained during moderate rainfalls. Undersized culverts at driveways and intersections typically caused the swales to store the watershed equivalent of a half inch of runoff (Jones, 1989).

Since the automobile was developed in the early years of this century, its use has been subsidized with public investment. Local governments spent millions of dollars to widen the old cobbled streets and repave them with asphalt. The federal government passed Federal Road Acts to construct and improve auto roads, support the formation and operation of state highway departments, and link state highways into national networks. In 1925 federal highway spending topped one billion dollars per year. In 1956 the federal government began the Interstate system, adding 41,000 miles of expressways and subsidizing the widening of local roads to feed automobiles onto the expressways. By the early 1970s paving was referred to as "the nation's biggest endowed business" (Sorvig, 1993). In the United States each year we are paving or repaving a half million acres of land (Ferguson, 1996).

City development was refitted to accommodate the car. Zoning codes imposed exclusive reliance on cars for daily transportation by segregating every detailed category of land use from every other and requiring homogeneous, low-density residential development across large areas. Pavements of new streets were more than 50 percent wider than the old ones to favor rapid, unobstructed traffic. The highways opened remote hinterlands to commercial and residential development. Citizens moved from farms to metropolitan areas, and from the central sections of cities to the suburbs. Businesses abandoned the old urban cores and relocated near suburban highway exits. In the suburbs, parking lots became essential adjuncts to stores and offices that had once fronted on city sidewalks. Downtown commercial districts lost their vitality as bulldozers leveled suburban hills and plains for shopping centers, industrial parks, and housing developments.

As land uses spread apart with only auto roads as connectors, more cars were needed to link them together again, and more asphalt and concrete were needed to maintain the connectors. Following the principle of "induced traffic," each highway built to alleviate congestion on an earlier existing road generated a larger aggregate amount of traffic for all roads. In most suburbs today it is physically dangerous to travel even short distances on routine business by any means other than a car.

HYDROLOGY OF CONTEMPORARY DEVELOPMENT

The new impervious pavements generated runoff. The new curbs and storm sewers accelerated it.

Figure 1.3 shows the amount of impervious cover that contemporary land uses produce. On the left of the chart, spread-out, large-lot residential areas cover 12 to 20 percent of the land with impervious surfaces. At the right of the chart are more intense commercial and industrial land uses, with shopping centers producing more than 90 percent impervious cover. In all types of land uses, most of the impervious cover is in pavements for automo-

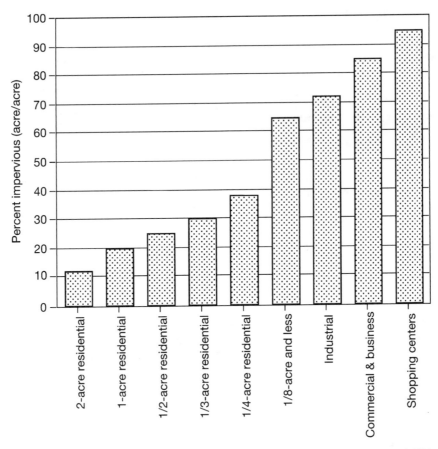

Figure 1.3 Impervious cover as a function of contemporary land use type (Arnold and Gibbons, 1996).

biles. In large-lot residential areas it is in driveways and local streets; in shopping centers it is in highways and parking lots.

Figure 1.4 shows the effects of impervious cover. Impervious roofs and pavements prevent natural absorbing, storing and balancing forces from operating. They seal over the organic mulch and the soil pores, depriving the soil of water and air. They deprive native plants of soil moisture. They deprive groundwater reservoirs of replenishment. They deprive the streams and wetlands of their sustenance. By deflecting water across the surface, they make floods bigger downstream, eroding as they go. Stressful flows spin off in cycles of flood, drought, erosion, and extinction.

The hydrograph in Figure 1.5 illustrates how the impervious effect shows up in stream flow. The figure's two curves contrast typical flows from vegetated soil and from impervious cover. The storm flow from vegetated soil is low and long because of the soil's absorption of rainfall; it is preceded and followed by base flows discharging slowly from landscape storage. In contrast, the storm flow from impervious cover is big and fast and is soon over; it is accompanied by little or no base flows. Impervious cover produces the worst of all possible hydrologic worlds: it enlarges storm flows into destructive bursts and, at the same time, it withdraws the resource of base flows. The marked bursts of storm flows in

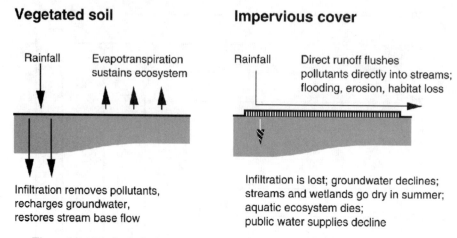

Figure 1.4 The hydrologic differences between vegetated soil and impervious cover.

Figure 1.2 reflect this effect: impervious roofs and pavements cover 50 percent of Peachtree Creek's watershed.

Urban flows surpass a channel's ability to maintain natural equilibrium. Amplified floods initiate incisive erosion. Bedscarps migrate upstream by undercutting the stream bed. High, vertical banks, unprotected by stable vegetation, collapse; sediment loads are high. Flood height and velocity are high and of short duration, while base-flow depths are shallow and uniform. The lowered channel drains groundwater from the floodplain, further reducing summer stream flow.

POLLUTION FROM CONTEMPORARY DEVELOPMENT

There is no chemically pure water in nature; in nature water quality can vary with climate, season, watershed mineralogy, and constituents carried in by precipitation. But every time

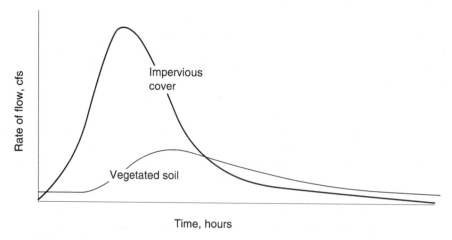

Figure 1.5 Contrasting storm flows from vegetated soil and impervious cover.

TABLE 1.1   Some of the Constituents of Stream Water

Constituent	Source in Nature	Role in Natural Ecosystem	Source of Urban Excess	Role of Excess
Sediment	Banks of meandering channels	Maintain stream profile and energy gradient; store nutrients	Construction sites; eroding stream banks	Abrade fish gills; carry excess nutrients and chemicals in adsorption; block sunlight; cover gravel bottom habitats
Organic Compounds	Decomposing organic matter	Store nutrients	Car oil; herbicides; pesticides; fertilizers	Deprive water of oxygen by decomposition
Nutrients	Decomposing organic matter	Support ecosystems	Organic compounds; organic litter; fertilizers; food waste; sewage	Unbalance ecosystem; produce algae blooms; deprive water of oxygen by decomposition
Trace Metals	Mineral weathering	Support ecosystems	Cars; construction materials; all kinds of foreign chemicals	Reduce resistance to disease; reduce reproductive capacity; alter behavior
Chloride	Mineral weathering	Support ecosystems	Pavement deicing salts	Sterilize soil and reduce biotic growth
Bacteria	Native animals	Participate in ecosystems	Pet animals; dumpsters; trash handling areas	Cause risk of disease
Oil	Decomposing organic matter	Store nutrients	Cars	Deoxygenate water

rain falls on an a city, it washes off oils, bacteria, litter, sediment, fertilizers, and foreign chemicals from streets, parking lots, lawns, dumpster pads, and metal roofs. Some 70 percent of the water pollution in the United States comes from "nonpoint" sources: the excess sediment, oils, and chemicals that runoff carries from eroding soil, parking lots, and intensely maintained lawns. Table 1.1 lists some common constituents and contrasts their roles in natural and urban watersheds.

In nature, sediment from gradually meandering stream banks builds point bars and aquatic habitats. In cities, excess sediment comes from eroding construction sites and from stream channels that are being torn apart by stressful stormwater flows. Additional particles come from glass, asphalt, stone, rubber, rust, and pavement fragments. In suspension, excess sediment makes water turbid, inhibiting plant growth and reducing species diversity. On settling to the bottom, excess sediment destroys spawning beds and the habitats of bottom-dwelling biota that depend on the interstices of sand and gravel particles for their habitat.

In nature, organic compounds come from the biodegradation of naturally occurring organic matter. In cities, excess organic compounds come from petroleum products, which are "organic" in the chemical sense. Among the "hot spots" for pollution in an urban watershed are gas stations, highway interchanges and heavily used commercial parking lots, where automotive oils and metals spill out. Excess organic decomposition deoxygenates water from within, weakening biotic communities in stream and lakes. Oil blankets the water surface, inhibiting reaeration. The loss of oxygen is reported in water quality studies by the measures of biological and chemical oxygen demand (BOD and COD).

In nature, nutrients such as nitrogen and phosphorus come from the organic detritus of riparian vegetation and support stream ecosystems. In cities, excess nutrients come from fertilizers on golf courses and other intensely maintained landscapes, septic tanks, sewer leaks and overflows, and leachates and debris from dumpsters and trash handling areas. In the United States, lawns cover a greater land area than any one agricultural crop, and many are maintained with excessive amounts of herbicides, pesticides, and fertilizers, which leach through the soil (Bormann, Balmori, and Geballe, 1993). The communities that adapt to an excess of nutrients are of low diversity, commonly dominated by a few species of algae. The decomposition of the algae's excess biomass deprives the water of oxygen.

On the whole, automobiles are the greatest source of pollutants in urban areas, after soil and streambank erosion. In streets, driveways, and parking lots, vehicles drop hydrocarbons, contained in oil, and metals produced by the wearing of brake pads and tires. Auto exhaust emissions pollute the air and, with precipitation, end up in the runoff. The large areas that streets cover, and the automobiles that use them, make streets major runoff and pollutant generators. Even stream bank erosion can be seen as an indirect product of automobiles, because it results mostly from the runoff from the pavements that cars require. The runoff and pollution from a contemporary city result not so much from the number of human beings, as from the lavish support given to their automobiles in land use and land development.

The roofs of buildings are relatively minor contributors to urban runoff and pollution because they tend to cover less area than pavements and usually have no way of producing pollution. In some residential areas, roofs drain onto the vegetated soil of lawns or foundation plantings, further reducing their runoff contribution. However, in industrial areas rooftops can be important because they are large and because some metal roofing is galvanized with zinc, which finds its way into runoff.

Impervious pavements can produce particularly high concentrations of some kinds of pollutants during a storm's "first flush" of runoff. When there is a period of a few days between rainstorms, oil, sediment, and debris accumulate on pavements. When a rain starts, the first flush of water carries off most of the pollutants; it has a high concentration as compared with runoff later in the storm (Figure 1.6). At these times—during small, frequent storms and the first runoff from large storms—constituent loads are not associated as much with quantity of runoff as with timing during the storm event.

TODAY'S URBAN STREAMS

Significant overall reduction of stream and wetland health, as measured by criteria such as pollutant loads, habitat quality, and aquatic species abundance and diversity, begins at 10 percent impervious coverage (Arnold and Gibbons, 1996). With impervious coverage of more than 30 percent, impacts become severe and degradation is almost unavoidable. Three broad categories have been established using simple numeric thresholds:

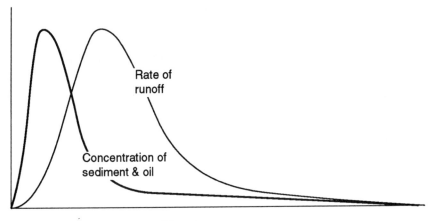

Time, hours

Figure 1.6   The "first flush" effect.

Impervious Coverage	Stream Health
< 10%	Protected
10 to < 30%	Impacted
≥ 30%	Degraded

Today urban streams, as a group, are probably the most disturbed and degraded in the country (Mikalsen, 1989). Contemporary urban streams are characterized by high bacterial density, high oxygen demand, high concentrations of solids and nutrients, high turbidity, and high concentrations of metals and organic compounds. The numbers and diversity of fish are low, and they decline as the impervious coverage in the drainage area increases.

When rain first falls on pavements, essentially all of it turns to runoff. It flushes the accumulated pollutants into streams. As the rain continues, growing volumes of runoff erode stream banks, destroying habitats and producing further sediment pollution. Bed materials shift; banks slough in; biota of all types are flushed out of chute-like channels. After the storm flow passes, base flow declines. Fish gasp for oxygen in the shallow, warm, sluggish water.

Hardly 7 percent of the land in the United States is classified as urban, but this is where 74 percent of the people live. They live among the most degraded streams in the country—and runoff problems are only one symptom of urban sprawl. Today's cities and suburbs suffer from traffic congestion, high energy consumption, air pollution, and daily dependence on automobile transportation.

THE POSSIBILITY OF RESTORATION

We are obligated to restore the mechanisms of the earth's self-maintaining balance. Runoff must be moderated, treated, and returned to its restorative path in the soil. To a large degree, it is possible to do so. Restoring environmental balance need not digress into clever mechanical plumbing or disruptively huge public works. It is a function, fundamentally, of

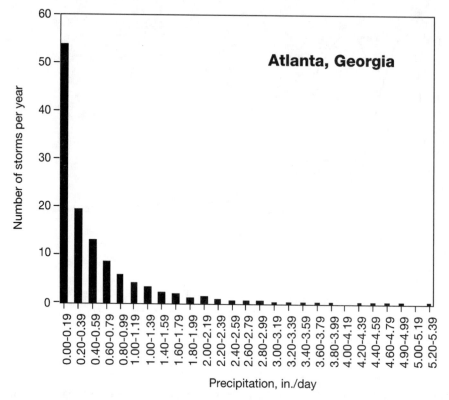

Figure 1.7   Frequency of storm events in Atlanta, Georgia (based on daily precipitation for the 40 years from 1950 through 1989).

how cities are built and used and the ways materials, earth forms, and plantings are used every day.

A given population can reduce its need for pavements by reducing its dependence on the automobile. Development that provides for nonautomotive transportation has compact mixes of land use, where many of people's everyday needs can be met within small distances, and safe, convenient paths for biking and walking among the various neighborhoods and land uses. When people are not using cars, there are no emissions and there is no demand for parking pavement when they arrive at their destinations.

Where paving is necessary, rainwater can be brought back into contact with the underlying soil through the use of permeable paving materials. Grass and crushed stone are old familiar "pavement" materials for the small traffic loads of parking stalls, residential drives, and pedestrian areas. Porous asphalt, porous concrete, and open-celled pavers appeared about 1970; together, they brought the possibility of hydrologic restoration to every street, golf cart path, bikeway, sidewalk, access lane, and parking lot.

Where impervious surfaces remain, conveying their runoff in vegetated swales and basins brings it back into contact with porous soil. Effective restoration need not be designed for huge volumes of water, because most of the rain that falls is in small, frequent storms (Figure 1.7). Big, flood-producing storms are destructive when they occur, but with decades between their occurrences, they contribute little to day-to-day environmental process. Treatment and infiltration of only a small runoff amount, when repeated for every small storm and the first flush of every large storm, restores most of a watershed's groundwater, base flow, and water quality.

Hydrologic restoration is not an economic or technological imposition upon nature. It is just nature. Nature wants to work. It evolves to work. If we can just stay out of nature's way, it will work. Stormwater management must reinitiate the kinds of long-term environmental processes that occurred before impervious surfaces were installed.

REFERENCES

Arnold, Chester L., and C. James Gibbons, 1996, Impervious Surface Coverage, the Emergence of a Key Environmental Indicator, *Journal of the American Planning Association* vol. 62, no. 2, pp. 247–258.

Bormann, F. Herbert, Diana Balmori, and Gordon T. Geballe, 1993, *Redesigning the American Lawn, A Search for Environmental Harmony*, New Haven: Yale University Press.

Driver, N. E., and G. D. Tasker, 1989, *Techniques for Estimation of Storm-runoff Loads, Volumes, and Selected Constituent Concentrations in Urban Watersheds in the United States*, Water-Resources Investigations Report 88-191, Denver: U.S. Geological Survey.

Ferguson, Bruce K., 1996, Preventing the Problems of Urban Runoff, *Renewable Resources Journal* vol. 13, no. 4, pp. 14–18.

Ferguson, Bruce K., and Philip W. Suckling, 1990, Changing Rainfall-Runoff Relationships in the Urbanizing Peachtree Creek Watershed, Atlanta, Georgia, *Water Resources Bulletin* vol. 26, no. 2, pp. 313–322.

Fisher, G. T., and B. G. Katz, 1988, *Urban Stormwater Runoff, Selected Background Information and Techniques for Problem Solving, with a Baltimore, Maryland, Case Study*, Water-Supply Paper 2347, Washington: U.S. Geological Survey.

Graf, W. L., 1975, Impact of Suburbanization on Fluvial Geomorphology, *Water Resources Research* vol. 11, no. 5, pp. 690–692.

Heaney, James P., and Wayne C. Huber, 1984, Nationwide Assessment of Urban Runoff Impact on Receiving Water Quality, *Water Resources Bulletin* vol. 20, no. 1, pp. 35–42.

Jones, D. Earl, 1989, Historic Perspectives, Questions and Future Directions, pp. 199–206 of *Multi-Objective River Corridor Planning*, Eve Gruntfest, editor, Madison, Wisconsin: Association of State Floodplain Managers.

Kunstler, James Howard, 1993, *The Geography of Nowhere, The Rise and Decline of America's Man-Made Landscape*, New York: Touchstone.

Latimer, James S., Eva J. Hoffman, Gerald Hoffman, James L. Fasching, and James G. Quinn, 1990, Sources of Petroleum Hydrocarbons in Urban Runoff, *Water, Air and Soil Pollution, WAPLAC* vol. 52, no. 1/2, pp. 1–21.

Leopold, Luna B., 1968, *Urban Hydrology for Land Planning*, Circular 554, Washington: U.S. Geological Survey.

Marsalek, Jiri, 1991, Pollutant Loads in Urban Stormwater: Review of Methods for Planning-level Estimates, *Water Resources Bulletin* vol. 27, no. 2, pp. 283–291.

Mikalsen, Ted, 1989, Factors Influencing the Quality of Urban Streams in Georgia and the Implications for Stream Management, pp. 135–138 of *Proceedings of 1989 Georgia Water Resources Conference*, Kathryn J. Hatcher, editor, Athens: University of Georgia Institute of Natural Resources.

Real Estate Research Corporation, 1974, *The Costs of Sprawl, Detailed Cost Analysis*, Washington: Council on Environmental Quality.

Schueler, Thomas R., 1995, *Site Planning for Urban Stream Protection*, Washington: Metropolitan Washington Council of Governments.

Seaburn, G. E., 1969, *Effects of Urban Development on Direct Runoff to East Meadow Brook, Nassau County, Long Island, New York*, Professional Paper 627-B, Washington: U.S. Geological Survey.

Simmons, Dale L., and Richard J. Reynolds, 1982, Effects of Urbanization on Base Flow of Selected South-Shore Streams, Long Island, New York, *Water Resources Bulletin* vol. 18, no. 9, pp. 797–805.

Sorvig, Kim, 1993, Porous Paving, *Landscape Architecture* vol. 83, no. 2, pp. 66–69.

Tasker, Gary D., and Nancy E. Driver, 1988, Nationwide Regression Models for Predicting Urban Runoff Water Quality at Unmonitored Sites, *Water Resources Bulletin* vol. 24, no. 5, pp. 1091–1101.

U.S. Environmental Protection Agency, 1990, National Pollutant Discharge Elimination System Permit Application Regulations for Storm Water Discharges, Final Rule, *Federal Register* vol. 55, no. 222, pp. 47990–48091.

CHAPTER 2

STORMWATER AND HUMAN EXPERIENCE

Stormwater management, like quality of human life, is influenced by, and influences in return, every detail of a site. It begins with an understanding of place. Every roof, pavement, channel, and basin is located in the midst of a living community. While we protect the biophysical environment, we must live in humane and prosperous cities. Those who propose to aid urban development must be aware of human and natural ecology, so they can perceive the forms and processes of a place and call them forth to make the land whole.

THE EXPERIENCE OF STORMWATER

People react to certain environments in terms of what those environments mean to them. The distinguishing characteristics of a place help to identify it; they make people aware of the place and its meaning.

The quality of experience is dynamic and creative (Koh, 1988). Over time, landscapes are produced and appreciated through participation and adaptation. Possessing and participating in an environment personalizes it, and changes and completes its meaning.

People derive particular satisfaction from places with natural characteristics. People seek natural settings both when they are harried or under pressure and in their everyday lives at home (Kaplan, 1982). People anchor themselves in their bioregions when they read the course of water through their communities (Thayer, 1994). A survey of residents along a Michigan creek that serves as a county "drain" found that although the creek's natural environment was distinctly unspectacular, it was appreciated for its "thereness" in the residential context (Kaplan, 1982). Even where the people were not doing much in or with this natural material, its mere presence mattered, as does, in other places, a small piece of open land or a tree outside a window. They expressed concern for their "nature amenity" and its future even though the drain was not everywhere in good condition.

But where drainage infrastructure is only a technical appendage to a community, local residents, recognizing culturally dysfunctional appliances as irrelevant or hazardous to their well-being, have had no motivation to learn about or to adapt to them. In Atlanta, almost half of the residents living near 15 small dry detention basins did not know that any basins existed in their immediate areas (Debo and Ruby, 1982). Those who were aware of the basins reported problems of silt and trash accumulation, concern for children's safety, mosquitoes, rats, snakes, and foul odors. None of them knew, correctly, who was responsible for maintenance of the basins; apparently no one was maintaining them at all. Those living downstream of the basins were not aware of any drainage benefits.

In order for a place where natural processes occur to be appreciated and maintained by people, human intention to care for the land must be evident (Nassauer, 1995). How a landscape shows that is tended by good citizens and good neighbors determines whether people interpret it as a natural place, an abandoned or neglected place, or a dumping ground. Neatness and order are signs that a place is under the care of people who have been there and return frequently. Familiar symbols of care include foreground mowing, structured paths, ornamental architectural details and "furniture," vividly ornamental trees and shrubs, orderly planting arrangements, and programs of maintenance by local residents. In the case of the Michigan "drain," the people preferred the creek when it represented a sense of order or domesticity, with well-defined edges, adjacent fine-textured lawns, and its flow well within bounds.

Where urban runoff occurs, a disturbance has taken place; mitigation and restoration are necessary. In the midst of a city, restoration of ecological processes depends on human protection and management. The characteristics of a place can make the processes through which hydrologic and ecological restoration takes place visible and comprehensible. Design is capable of revealing and integrating.

One measure of human response to a place is property value. Stormwater ponds and wetlands have raised property values where they have been integrated into the visage and life of their communities (U.S. Environmental Protection Agency, 1995). In Alexandria, Virginia, condominiums that fronted on a stormwater lake sold at a $7,500 premium. In St. Petersburg, Florida, apartments facing small ponds rented at a $15 per month premium; for those facing a large pond the premium was $35. In Wichita, Kansas, lots lining wetlands that had been adapted for stormwater treatment sold for as much as 50 percent more than comparable lots with no water view.

Similarly, in Atlanta, residents living near small lakes believed that the lakes raised the values of their homes; their presence was a positive factor in the decision to buy a home (Debo, 1977). Residents expressed willingness to pay for lake maintenance if they had access to and could utilize the recreational facilities located on the lake shores. Those living near larger lakes (18 to 48 acres) with more intensely developed recreational facilities, such as beaches and tennis courts, had a more decidedly positive attitude about the lakes and their potential expenses; this attitude was held even by people not living immediately adjacent to the lakes.

In Illinois, Emmerling-DiNovo (1995) distinguished between the effects of dry and wet detention basins. She asked residents of subdivisions containing basins of 2 to 12 acres surrounded by grass to estimate the impact of the basins on residential images and lot values. They perceived lots adjacent to wet basins to be the most valuable, adding an average of 22 percent to the value of similar nonadjacent lots; assessed lot values, adjusted for size, were in fact 10 percent higher than the average for all lots in the same subdivisions. They perceived lots adjacent to dry basins to be the least valuable; dry basins added little or no value to adjacent properties. When presented with six locational options, the most favored location for all respondents was adjacent to a wet basin and the least favored was adjacent to a dry basin.

In the Village Homes community in Davis, California, the distinctively "open" drainage system is close to the lives of the residents (Corbett, 1981; Thayer, 1977; Thayer and Westbrook, 1989). Meandering swales are near the back doors of all the houses (Figure 2.1). Narrow street pavements drain into the swales through curb notches; roofs drain onto vegetated soils that slope toward the swales. In the swales, numerous wooden and rock check dams, 6 to 8 inches high, form pools where runoff percolates into the soil. When each pool is full, the runoff spills into the next. The bike paths that parallel the swales link playgrounds and open spaces.

Figure 2.1   Open swale managed by residents in Village Homes, Davis, California.

The residents in each cluster of eight homes designed, implemented, and continue to maintain and adjust the materials and plantings in the swales and open spaces. The result is astonishing diversity of space and purpose. The pond floors are variously surfaced with grass, clover, sand, gravel, and river cobbles. Small wooden bridges span the swales for foot and bike traffic; stepping stones lead to individual homes (all were paid for by savings from not constructing buried pipes). Riparian trees and subirrigated fruit and nut orchards line the swales; individual home owners experiment with water-loving border plants. The riparian mini-ecosystem visibly supports native birds, mammals, and butterflies in linked moisture zones of diverse vegetation.

Village Homes' swales reduce peak flows while producing the ephemeral sounds and smells of trickling water and damp earth. The spirit of caring for growing things has fostered community arrangements of all kinds. The drainage system is part of a cultural and ecological structure that binds rainfall, soil, plants, wildlife, human senses, human work, and human knowledge through direct interaction. What the system looks like, how it functions ecologically and socially, and what it symbolizes in the way of stewardship, are congruent.

RESTORATIVE COMMUNITIES

Knowledgeable design of urban development solves the problem of runoff at the source—in the land uses where people live, where pollutants are first generated, and where urban rainwater first touches the ground. The solution is embedded in soil, vegetation, transportation, land use, and the human way of life.

Patterns of Community Development

Every aspect of land use allocation, site planning, and detailed design can be brought to bear on stormwater restoration. An overall measure of success is the amount of impervious surface produced by a given unit of development.

In contemporary American development, cars are the principal generators of impervious surfaces. Wherever a car goes, it leaves a parking space of 300 to 350 square feet, travels in a lane occupying 2,000 square feet or more, and arrives at a second parking space of still another 300 to 350 square feet. Automotive pavements occupy more than half the impervious surface in residential developments; in commercial areas they occupy 80 percent of all the land. Although cities are supposed to be the homes of people, they are built predominantly for cars. In large part, the effort to limit impervious cover is aimed at pavements and the automobiles that require them.

Both dispersed, sprawling, low-density, auto-oriented land use, on one hand, and clustered and infill developments, on the other, produce impervious roofs and pavements, but they do so in different ways and to different degrees (Figure 2.2).

In dispersed land use (such as large-lot single-family residential) the total and per capita quantities of impervious surfaces are high for a given number of residents, because autos are necessary to transport things and the area paved to connect the dispersed buildings is large.

However, cluster development concentrates a given quantity of land use on only a portion of the available land by using reduced lot sizes, reduced street lengths, reduced setbacks, and sharing of driveways and parking bays. The decreased infrastructure reduces

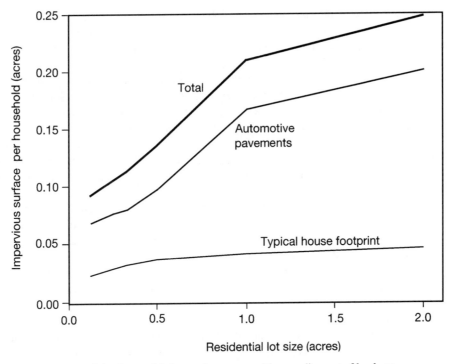

Figure 2.2   Increasing impervious cover with spreading out of land use.

development cost. Close-set buildings define public places by enclosing streets spatially (Newman, 1995). Clustered developments generate local concentrations of runoff and pollutants from their concentrated pavements, autos, and people. But clustering uses a relatively small amount of pavement to support a given unit of development while leaving other large areas pristine, so the total runoff and pollution from a site or a region are lower than they would be with dispersed development (Table 2.1).

Infill development and redevelopment in and near previously developed places is longterm clustering on a regional scale. Blending different, but related, land uses into compact

Table 2.1   Hydrologic Effects of High and Low Densities of Land Use

	High density (clustered & infill) land use	Low density (dispersed) land use
Land consumption for a given population	Low	High
Total quantity of runoff and pollution for a given population	Low	High
Local concentration of runoff and pollution	High	Low
On-site absorption of runoff and pollution	Low	High

neighborhoods reduces dependence on automobile transportation and the consequent pavements, emissions, and runoff. Although infilling results in locally high amounts of impervious cover in the built-up areas, it reduces sprawl, impervious cover, and auto usage in the region as a whole.

Connecting nearby related land uses with sidewalks and bikeways eliminates some of the demand for pavements. Walking and biking require only small paved surfaces: the traveling lanes for bikes are only half as wide as those for cars, and 15 bikes can be parked in the space required for one car. By combining exercise with commuting, walking and biking save energy, reduce traffic congestion, and contribute to individual health. These means of transportation are low in cost. They are distinctively compatible with the quietness and safety of residential neighborhoods. Walking and biking help to eliminate runoff and pollution by never generating them in the first place.

Suburban parking has, in fact, been oversupplied nationally even in auto-oriented developments and deserves to be reduced. Ordinances commonly require a ratio of 4 parking spaces per 1,000 square feet of office space; this amount of parking occupies from 30 to 70 percent of the site. But a peak ratio of 2.8 per 1,000 is the utilization found by actual observation (Wilson, 1995). Excess unutilized parking adds no value to a development. It raises construction and maintenance costs, and contributes to the "heat island" effect that elevates temperatures and demands energy for air conditioning. Parking that is unused by commuters is "used" by rainfall every time it rains.

Similarly, street pavement width should be limited to no more than that needed for the function of each specific street. Categorizing streets according to the fronting land use and the amount of traffic furnishes a structure for street standards that fit the needs of the land use for transportation and parking, as well as the needs of the street for traffic flow and runoff treatment and drainage. Street design can respond with site-specific allocations of curbing, swales, on-street parking, and lane width (Georgia Environmental Protection Division, 1997; Richman and Associates, 1997; Schueler, 1995).

Patterns of Construction Details

Within a given land use, you can lay your hands on the details of construction materials to bring restorative processes to every inch of inhabited places. Disconnect impervious surfaces; grade them to drain onto vegetated soil. Turn impermeable surfaces into permeable ones, and inert ones into living ones. Use native plant communities that need no petrochemical lawn care. Nurture permeable soils with compost, mulches and densely rooted plants. Preserve preexisting riparian vegetation along stream and swale corridors. Allow self-maintaining biotic systems to take over.

Overflow parking on grass is a casually civilized practice, where the parking arrangement is organized along a paved traveling lane (Figure 2.3). Grass maintains its health, appearance and permeability at parking frequencies up to about once a week. Wherever similarly modest pedestrian and vehicular traffic loads occur, living vegetation maintains open, absorptive soil.

For slightly heavier or more frequent traffic loads, such as borne by pedestrian plazas, residential driveways, and the parking stalls of parking lots, crushed stone aggregate is an almost maintenance-free material. The crunching sound of a stone drive or walk has associations with old estates and refined civic facilities. Open-graded stone sizes such as #67 and #89, which leave open voids between uniformly sized aggregate particles, are highly

Figure 2.3   Overflow parking on grass during a busy summer Saturday afternoon at the Fox Chapel Golf Club in Pittsburgh.

permeable. Bollards, wheel stops, paved traveling lanes, or arrangements of plantings can organize parking where painted lines are not possible. Crushed stone is cost-competitive with almost any other paving material.

For areas that must support heavier traffic, structural porous pavements must be selected and designed for site-specific soils, slopes, drainage, and traffic loads, as are any other pavements.

Parking areas, especially, deserve porous treatment. At increasing distances from building entrances, parking spaces are used less often; in many existing parking lots the most distant spaces are literally never used. To make parking pavements permeable is to attack one of the least necessary sources of urban runoff.

Both grass and aggregate can be reinforced with open-celled pavers (Nichols, 1995; Sipes and Roberts, 1994; Southerland, 1984). Figure 2.4 shows grid concrete pavers with cells filled with porous aggregate. At the Westfarms Mall in Farmington, Connecticut, 4.7 acres of overflow parking are in turf reinforced with a grid of recycled plastic, fulfilling a local quota for "green space" while meeting the functional needs of the land use (Thompson, 1996).

Porous asphalt and concrete are capable of meeting the structural needs of heavy loads (Florida Concrete and Products Association, n. d.; Thelen and Howe, 1978). In both, the distinctive surface layer is constructed with open-graded aggregate. Beneath the surface, an open-graded stone base holds water until it infiltrates the underlying soil. The first installation of porous asphalt was a parking lot at the University of Delaware Visitors Center in Newark (Bachtle, 1974); as of the summer of 1996 it was still in structurally perfect condition after decades of use and exposure. In many respects, porous asphalt and concrete are

Figure 2.4 Open-celled pavers filled with aggregate at an office building on Longboat Key, Florida.

superior to their impervious cousins. Their surfaces are better drained, so in wet weather they give safer traction. Visibility is better, because there is no layer of water over the pavement and vehicles do not kick up plumes of mist from their wheels. Because of its safety advantages, the distinctive surface layer of porous asphalt has been used in every type of pavement from residential driveways and parking lots to Interstate highways.

The variety of materials available today bring permeable pavements to every type of urban place, amid varying functional, structural, aesthetic, and cost constraints. Impermeable materials should be constructed only as special cases, in response to site-specific constraints of soil, slope or traffic load. In a restorative community, the use of permeable materials, visibly distinctive with their patterns of open voids and, in some cases, their living vegetation, should be a way of life. Wherever we go, we should be conscious of the careful return of rainfall to the soil.

The design of Cardinal Ridge in Medford, New Jersey, pursued aggressively every detail of layout and construction to preserve soil permeability. Cardinal Ridge was the first development constructed under Ian McHarg's environmental study for Medford Township (Juneja, 1974). Houses were clustered on small lots to reduce the lengths of local access streets, and the pavements of connecting roads were minimized in width. Driveways are of crushed stone. Residential patios are on wooden decks. Residential walkways are of wood mulch, well spaced paving stones, and stone aggregate. The character of Cardinal Ridge, which is a single-family subdivision far from the town center, is of dense shading under the essentially undisturbed native forest canopy. In Cardinal Ridge one seems to drive through and live in a forest, with the intact, densely shaded forest floor and native vegetation always adjacent to homes and drives.

Figure 2.5   Crushed stone parking stalls in Medford Village, New Jersey; the traveling lane is of ordinary asphalt.

A different character is found in the town center of Medford Village, where closely situated historic structures are connected by pedestrian sidewalks. McHarg's restorative vision is carried out where parking lots crowding the alleys and back lots are of crushed stone, or combinations of crushed stone and asphalt (Figure 2.5). Here porous materials, bringing the natural processes of rain, soil, and life together, have found their way into an urbane and densely inhabited place, and it is only appropriate that they do so. They express a careful way of life. The community of Medford Village, with its municipal offices, storefront shops, and upstairs apartments, is an active partner in the soils and aquifers that underlie it, the trees that shade it, and the streams that drain around it. People here are aware of the place where they live, and they take care of it.

RESTORATIVE DRAINAGE

There will always be some measure of runoff from roofs and unavoidable impervious pavements. To compensate, to some degree, for the impervious cover that is built, by using restorative drainage systems is to complete the community.

Drainage restoration starts near the sources of runoff in the midst of the community. Beginning at the source uses the full capacity of a site's vegetation and soils for restoration and the full capacity of the site's residents for perception and participation. Swales and basins should be small and numerous; they should be located at every downspout, every curb cut, and every small area of pavement. They should continue downstream, each stage bringing water again into contact with vegetation and soil and adding some improvement to the quantity and quality of the runoff.

Swales

In vegetated swales, unlike paved gutters or structural pipes, vegetated soil infiltrates and stores rainfall, treats it, and discharges it gradually to streams weeks after storms are past.

In the Grogan's Mill community near Houston, hydrology organizes and binds the layout of streets, building clusters, and open spaces (Girling, 1994; McHarg and Sutton, 1975; Sutton, 1974). Grogan's Mill was the 1,900 acre first phase of The Woodlands, a large development where construction began in the early 1970s. The site's floodplains are broad and flat; one third of the site is within the 100-year floodplain. In setting environmental design criteria for the development, Ian McHarg saw water as an integrating process that explains the nature of the site through its flows over and in the ground and its effects on soil and vegetation.

Into each of Grogan's Mill's residential clusters, wide shallow swales lined with native vegetation extend the natural drainage system. Driveway bridges cross the swales to reach private lots, where landscaping is limited mostly to wooden decks under the woodland canopy. Ponds and check dams infiltrate runoff into permeable soils. Overflows from the swales pass into the natural stream system, which is preserved in wooded and recreational open spaces for flood storage and infiltration. Paved trails paralleling the drainage corridors make swales, streams, and basins apparent everywhere in the wooded community. As an alternative to building structural channels, the open drainage system saved more than $14 million.

Strawberry Creek Park in Berkeley, California, represented a greater challenge to the nurture of the values of flowing streams, because it required the "daylighting" of a creek in an old urban area, bringing a previously culverted stream back into the life of the city in the light and air (Strawberry Creek Park, 1995). The four-acre site had been an abandoned railroad yard surrounded by vacant industrial buildings, frequented by drug dealers and feared by neighbors; the creek was in a large culvert below. In 1984 the city remade the place into a neighborhood park with cooperative funding from the California Land and Water Conservation Fund (Figure 2.6). Recreation courts, play areas, and simple sitting spots and open spaces were chosen by neighborhood design participants and are actively used by them today.

The park site was regraded for surface drainage toward the central creek channel. Concrete paving was demolished and recycled as riprap to stabilize the creek banks, giving the sharply meandering channel the character of a rugged mountain stream. In the center of the park picnic tables are adjacent to a bike- and- footbridge that crosses the stream amid native riparian plants. The flow is rapid during winter rains; it falls in white aerating steps

Figure 2.6 Strawberry Creek Park, Berkeley, California.

over boulderlike fragments of recycled concrete. The water's diverse noises come from many spots in the meandering, falling channel; they make a marked and welcome contrast to the sounds of vehicles outside the park. The water is gray with urban pollution from upstream, but its brief aeration and exposure to sunlight in the park add one increment to its quality.

Shortly after the park was completed, nearby buildings began to be occupied by senior housing, cohousing, day care, offices, shops, an adult school, and an employment center where disadvantaged youths are trained to maintain the park. Local materials, participatory design, local caretakers, and drawing people and nature back to the neighborhood, weave the park into its context.

Basins and Dams

Locating stormwater basins centrally and integrally within a community makes them visible and accessible. Meaningful amenities are thereby created; nearby residents are likely to act as overseers of basin maintenance and safety.

The contouring of a basin and its surroundings creates the depth and elevation zones to which various categories of biota are adapted. Fish require open water with limited aquatic vegetation; they need various bottom materials, depths, and levels of shading to diversify food and cover. Waterfowl require interspersed spits, inlets, islands, and open water; extensive edges and shallows are their feeding places; diverse shrubs and grasses make places for nesting, loafing, and cover. On the other hand, to suppress mosquitoes and midges, ponds should be free of extensive shallow zones where still water and thick plant beds would provide cover, and free of excess eutrophying organic matter (Mackenthun, Ingram, and Porges, 1964, p. 97–100).

Dams and their outlets present design alternatives that have too seldom been exploited for the benefit of human communities (Urbonas et al., 1985; U.S. Bureau of Reclamation, 1974). Parts of dams can be widened, additional fill placed, and slopes made gentle, long, and rounded, merging into surrounding landforms; as long as material is only added to a cross-section, not removed, a dam is not weakened. A weir or flume conducting water over the face of a dam allows spilling water to be seen and heard.

Figure 2.7 shows a highly architectural pond in the University Place development in Charlotte, North Carolina. This is an intensely urban stormwater reservoir. Restaurants, shops, hotels, and offices in the surrounding buildings front directly on the walkways and plazas that line the pond. The fountain aerates the stormwater that passes through the pond. The pond's edges can be criticized because they are ecologically simple and homogeneous. However, from a human viewpoint, the walkway edges are well marked and well maintained; the plane of water is the central open space that unifies the place. The safe experience here indicates that fencing may not be necessary around basins where edges are visible.

Figure 2.8 shows an attempt to bring restorative natural processes to an old urban pond at Lake Eola in Orlando, Florida. The lake is in a heavily used downtown park. Native plants such as cypress and yellow taro were planted in the pond's shallow edges at the same time that the park facilities were rehabilitated and stormwater treatment devices were retrofitted into the upstream storm sewers. All was done in a combined effort to improve water quality and bring the park more closely to the life of the city. That native egrets are willing to fish in the water while the park is full of people during lunch hour is a sign of at least partial success.

Figure 2.7   University Place, Charlotte, North Carolina.

Figure 2.8   Lake Eola, Orlando, Florida.

Human Safety

Informal surveys have shown that where people have been hurt around drainage facilities, more than half of the places have been characterized by concentrated quantities of fast-moving water, particularly by constriction of flow at the mouths of culverts.

An example is an ephemeral stream in State College, Pennsylvania, that drains a residential subdivision and then passes through a park. Shortly after a large storm, a four-year-old girl walking with her family in the park slid down wet, slippery grass into the stream and became trapped against a trash rack at the entrance to a culvert. The trash rack was properly sloped at a 45 degree angle, which is supposed to allow entering water to push objects up and out of the way. But there was so much water entering the culvert that it was pushing down on the child as it went through the trash rack. Two men who tried to pull her away failed, and she drowned.

Another example is a dry detention basin in East Brunswick, New Jersey, located between two residential subdivisions. The basin discharges through a 21-inch culvert into a series of storm sewers three miles long. During a flood two 15-year-old boys watched the water rise, in what was normally a weedy field, to become a lake the size of a football field, 6 to 10 feet deep. Early in the afternoon a neighbor saw them throwing stones into the violent vortex over the outlet and warned them away. But in the evening both boys were missing from their homes. The dog that had been with them was found bound to a tree near the basin. The next morning searchers found their bodies, one a mile downstream in a 40-inch section of storm sewer, the other two miles farther down, where the pipes discharged into a creek. They were both fully clothed. Perhaps one of them fell in while watching the whirlpool, and then the other fell in while trying to help. The suction of the vortex and the enclosure of the pipes allowed them no escape once they had slipped on the wet, grassy, steep-sided dam.

These examples of concentrated, enclosed flow argue that for safety, as much as for participation and restoration, conveyances should be as open and freely flowing as possible. Flow should be slow, shallow, and free of constrictions. At basin outlets, broad, open weirs are safer than anything involving the mouth of a pipe.

The places where people have been hurt in still ponded water have commonly been characterized by steep, slippery sides of clay or wet grass, deep water, and nothing to hold onto. If you were going to build a trap for people to fall into and drown, that is the way you would design it.

To make a pond safe, take each of those designed-in hazards and do the opposite. Make it unlikely that persons will fall in and, if they do fall in, will be able to pull themselves out again. A basin should be open, visible, and accessible, so that people can appreciate any hazard, conduct themselves carefully, monitor others, and, when necessary, provide rescue efforts quickly. Mark the approaches to water with changes in ground cover. Make side slopes gentle and edges shallow. Make the ground surface (both around the pond and on the bottom) rough, using pavements or ground covers. Break up water with islands and boulders; put solid objects for holding onto in the flood area.

The following two court cases (Kozlowski, 1985) illustrate some of the factors that have been considered in deciding liability for pond safety hazards.

In Illinois (*Cope v. Doe*, 1984) a seven-year-old boy drowned in a detention pond in an apartment complex. The Illinois court found the developer not liable for the boy's death. The pond was located only 100 yards from the apartment buildings and was clearly visible. The court considered it an "open and obvious danger," which children of age to be allowed at large should be expected to appreciate and avoid. On the day of the accident the pond was one-third covered with ice; open water could easily be seen. Children who had grown up in Illinois for seven years were expected to understand the hazard of thin ice. Nevertheless, the boy went on the ice and fell through while kicking pieces of wood. The apartment complex had been designed to attract families with young children. It included a children's playground, a swimming pool, tennis courts, and other recreational features. No precautions were taken to prevent children from going near the pond other than the manager's warning parents to keep children away from it. The primary purpose of the pond was flood control; although children were not prohibited from going near it, they were not invited to it for swimming or other recreational purposes as they were to the swimming pool.

In Louisiana (*Guillot v. Fisherman's Paradise Inc.*, 1983) a two-year-old boy visiting a vacation home drowned in a nearby marina's sewage oxidation pond. The Louisiana court found the marina's developer liable for the child's death. The pond was located 100 yards from the house where the parents were staying but was completely unmarked; none of the adults was aware of its presence. There was no closed border around the pond; a partial wooden fence was only for appearance. The pond had very steep side slopes to the its full depth of three feet. The adults had warned the child about a large nearby lake, which was the only water hazard of which they were aware. The little boy, who had never been allowed to go to the lake except in the company of adults, was considered an obedient child. He had wandered away from the house yard; within 10 minutes the adults noticed his absence and began to search for him. The Louisiana court did not expect parents ordinarily to keep watch over a young child every minute, or to keep a child in chains to prevent him from ever wandering. The boy had apparently wandered into the pond, which was covered with light green algae that made the presence of water less apparent and contained some pieces of bright-colored litter that may have been attractive to a young child. Almost an hour had passed before the parents discovered the child's body floating face down in the pond.

Hazards to safety can be recognized during design and eliminated. Where the presence of water is apparent, people are expected to, and usually do, act as if it is as dangerous as it looks. Facilities can be designed to make the falling and trapping of persons unlikely and to make it possible to get to them when trouble occurs.

REFERENCES

Arendt, Randall G., 1996, *Conservation Design for Subdivisions: A Practical Guide to Creating Open Space Networks*, Washington: Island Press.

Bachtle, Edward R., 1974, The Rise of Porous Paving, *Landscape Architecture* vol. 64, no. 5, p. 385-387.

Bormann, F. Herbert, Diana Balmori, and Gordon T. Geballe, 1993, *Redesigning the American Lawn; A Search for Environmental Harmony*, New Haven: Yale University Press.

Calabria, Tamara Graham, 1995, *The Representation of Stormwater Management in Design: Toward an Ecological Aesthetic*, MLA thesis, Athens: University of Georgia.

Corbett, Michael, 1981, *A Better Place to Live*, Emmaus: Rodale Press.

Debo, Thomas N., 1977, Man-Made Lakes: Attitude Surveys of Their Value in Residential Areas, *Water Resources Bulletin* vol. 13, no. 4, pp. 665–676.

Debo, Thomas N., and Holly Ruby, 1982, Detention Basins: An Urban Experience, *Public Works*, January, pp. 42–43, 93.

Emmerling-DiNovo, Carol, 1995, Stormwater Detention Basins and Residential Locational Decisions, *Water Resources Bulletin* vol. 31, no. 3, pp. 515–521.

Evans, David, and Associates, Inc., 1992, *What Needs to Be Done to Promote Bicycling and Walking?* National Bicycling and Walking Study, Case Study No. 3, Washington: U.S. Federal Highway Administration.

Ewing, Reid, 1995, *Best Development Practices: Doing the Right Thing and Making Money at the Same Time*, Fort Lauderdale: Florida Atlantic University/Florida International University Joint Center for Environmental and Urban Problems.

Ferguson, Bruce K., 1991, Taking Advantage of Stormwater Control Basins in Urban Landscapes, *Journal of Soil and Water Conservation* vol. 46, no. 2, pp. 100–103.

Ferguson, Bruce K., 1994, A Re-evaluation of Minimum Slope Standards, pp. 108–115 of *Conference Proceedings, CELA 94*, James D. Clark, editor, Washington: Landscape Architecture Foundation.

Florida Concrete and Products Association, n. d., *Pervious Pavement Manual*, Orlando: Florida Concrete and Products Association.

Georgia Environmental Protection Division, 1997, *Potential Development to Protect Runoff Quality*, Atlanta: Georgia Department of Natural Resources, Environmental Protection Division.

Girling, Cynthia L., 1994, The Marketing of Recreation and Nature: The Woodlands, Texas Revisited, in *Public Lands/scapes, Proceedings of the 1993 Conference of the Council of Educators in Landscape Architecture*, Robert G. Ribe, Robert Z. Melnick, and Kerry Ken Cairn, editors, pp. 43–56, Washington: Landscape Architecture Foundation.

Jones, Jonathan, and D. Earl Jones, 1982, Interfacing Considerations in Urban Detention Ponding, in *Proceedings of the Conference on Stormwater Detention Facilities*, New York: American Society of Civil Engineers.

Jones, Jonathan, and D. Earl Jones, 1984, Essential Urban Detention Ponding Considerations, *Journal of Water Resources Planning and Management* (American Society of Civil Engineers) vol. 110, no. 4, pp. 418–433.

Juneja, Narendra, 1974, *Medford: Performance Requirements for the Maintenance of Social Values Represented by the Natural Environment of Medford Township, N.J.*, Philadelphia: University of Pennsylvania Department of Landscape Architecture and Regional Planning, Center for Ecological Research in Planning and Design.

Kaplan, Rachel, 1982, The Green Experience, pages 186-193 of *Humanscape, Environments for People*, Stephen Kaplan and Rachel Kaplan, editors, pp. 186–193, Ann Arbor: Ulrich's Books.

Koh, Jusuck, 1988, An Ecological Aesthetic, *Landscape Journal* vol. 7, no. 2, pp. 177–191.

Komanoff, Charles, and Cora Roelofs, 1993, *The Environmental Benefits of Cycling and Walking*, National Bicycling and Walking Study, Case Study No. 15, Washington: U.S. Federal Highway Administration.

Konrad, Christopher P., Bruce W. Jensen, Stephen J. Burges and Lorin E. Reinelt, 1995, *On-Site Residential Stormwater Management Alternatives*, Seattle: University of Washington Center for Urban Water Resources Management.

Kozlowski, James C., 1985, Retention Pond Drowning Case Studies, *Parks & Recreation*, March, pp. 38–44, 95.

Leccese, Michael, 1996, Pedestrian Friendly, *Landscape Architecture* vol. 86, no. 9, pp. 36–41.

Leccese, Michael, 1997, Cleansing Art, *Landscape Architecture* vol. 87, no. 1, pp. 70–76 and 130.

Litton, R. Burton, Robert J. Tetlow, Jens Sorensen, and Russell A. Beatty, 1974, *Water and Landscape: An Aesthetic Overview of the Role of Water in the Landscape*, Port Washington, N. Y.: Water Information Center.

Mackenthun, Kenneth M., William Marcus Ingram, and Ralph Porges, 1964, *Limnological Aspects of Recreational Lakes*, Publication 1167, Washington: U.S. Public Health Service.

McHarg, Ian L., and Jonathan Sutton, 1975, Ecological Plumbing for the Texas Coastal Plain, *Landscape Architecture* vol. 65, no. 1, pp. 78–89.

Nassauer, Joan Iverson, 1995, Messy Ecosystems, Orderly Frames, *Landscape Journal* vol. 14, no. 2, pp. 161–170.

Newman, Oscar, 1995, Defensible Space: A New Physical Planning Tool for Urban Revitalization, *Journal of the American Planning Association* vol. 61, no. 2, pp. 149ñ155.

Nichols, David, 1995, Comparing Grass Pavers, *Landscape Architecture* vol. 85, no. 5, pp. 26–27.

Real Estate Research Corporation, 1974, *The Costs of Sprawl: Detailed Cost Analysis*, Washington: Council on Environmental Quality.

Richman, Tom, and Associates, 1997, *Start at the Source: Residential Site Planning and Design Guidance Manual for Stormwater Quality Protection*, Oakland, California: Bay Area Stormwater Management Agencies Association.

Schueler, Thomas R., 1995, *Site Planning for Urban Stream Protection*, Washington: Metropolitan Washington Council of Governments.

Sipes, James L, and John Mack Roberts, 1994, Grass Paving Systems, *Landscape Architecture* vol. 84, no. 6, pp. 31–33.

Sorvig, Kim, 1993, Porous Paving, *Landscape Architecture* vol. 83, no. 2, pp. 66ñ69.

Southerland, Robert J., 1984, Concrete Grid Pavers, *Landscape Architecture* vol. 74, no. 2, pp. 97–99.

Strawberry Creek Park, 1995, *Landscape Architecture* vol. 85, no. 11, pp. 46–47.

Sutton, John, 1974, Hydrological Balancing Act on a Texas New Town Site, *Landscape Architecture* vol. 75, October, pp. 394–395.

Taylor, Albert D., 1935, Design and Construction of Small Earth Dams, *Landscape Architecture* vol. 25, no. 3, pp. 138–157.

Thayer, Robert L., 1977, Designing an Experimental Solar Community, *Landscape Architecture* vol. 67, no. 3, pp. 223–228.

Thayer, Robert L., 1994, *Gray World, Green Heart: Technology, Nature, and the Sustainable Landscape*, New York: John Wiley & Sons.

Thayer, Robert L., and Tricia Westbrook, 1989, Open Drainage Systems for Residential Communities: Case Studies from California's Central Valley, in *CELA '89: Proceedings* (Proceedings of the 1989 Conference of Council of Educators in Landscape Architecture), Sara Katherine Williams and Robert R. Grist, editors, pp. 152–160, Washington: Landscape Architecture Foundation.

Thelen, Edmund, and L. Fielding Howe, 1978, *Porous Pavement*, Philadelphia: Franklin Institute Press.

Thompson, J. William, 1996, Let That Soak In, *Landscape Architecture* vol. 86, no. 11, pp. 60–67.

Tourbier, J. T., 1994, Open Space Through Stormwater Management: Helping to Structure Growth on the Urban Fringe, *Journal of Soil and Water Conservation* vol. 49, no. 1, pp. 14–21.

Urbonas, Barnabas R., and others, 1985, *Stormwater Detention Outlet Control Structures*, Report of the Task Committee on the Design of Outlet Control Structures, New York: American Society of Civil Engineers.

U.S. Bureau of Reclamation, 1974, *Design of Small Dams*, second edition, Washington: U.S. Bureau of Reclamation.

U.S. Environmental Protection Agency, 1995, *Economic Benefits of Runoff Controls*, EPA 841-S-95-002, Washington: U.S. Environmental Protection Agency.

Whyte, William H., 1964, *Cluster Development*, New York: American Conservation Foundation.

Wilson, Richard S., 1995, Suburban Parking Requirements and the Shaping of Suburbia, A Tacit Policy for Automobile Use and Sprawl, *Journal of the American Planning Association* vol. 61, no. 1, pp. 29–42.

CHAPTER 3

INTRODUCTION TO HYDROLOGIC CONCEPTS

Before using hydrology in applications, you have to be comfortable with the basic physical concepts that underlie it. At its roots hydrology is fundamentally simple. It amounts to moving volumes of water over time. It is underlain by—indeed, almost entirely composed of—the basic concepts of volume and rate. If you are versed in those basic concepts, you can be comfortable with all the ways we look at water in this book. To see how simple the foundations of hydrology are, let us review everything that most people know about volumes of water. Equivalencies between some units of measure are listed in Table 3.1.

Table 3.1   Equivalencies among some units of measure.

Amount	Equivalent Amount
1 meter	3.281 ft
1 cubic meter	35.32 cf
1 liter	0.03532 cf
1 gallon	0.1337 cf
1 acre-foot (af)	43,560 cf
1 acre-inch (ac-in.)	$\frac{1}{12}$ af
1 hectare	2.471 ac
1 square kilometer	247.1 ac
1 meter per second	3.281 fps
1 cubic meter per second	35.32 cfs
1 gram (g)	0.002205 lb
1 milligram (mg)	0.000002205 lb
1 kilogram (kg)	2.205 lb

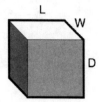

Figure 3.1   Volume = LWD.

VOLUME AND TIME

A static volume, such as that of a cube, is equal to the product of the three sides: length L, width W, and depth D (Figure 3.1):

$$Volume = LWD$$

Because volume is the product of three dimensions, its units are cubic, such as cubic feet (cf). Because LW equals area A, we can also say that volume is equal to the product of area and depth (Figure 3.2):

$$Volume = AD$$

Thus we could analyze a volume of water from the side (depth), from the top (area) or as a total volume. For example, the depth D might be the depth, in inches, of rainfall in a given period of time; the total volume of water that the rain adds to a watershed is equal to the depth of rainfall times the area of the watershed. Or the area A might be the area available in a site for locating a stormwater storage reservoir, in acres; if a given volume must be stored, then the required depth can be derived by solving the preceding equation for D.

Velocity (Figure 3.3) is defined as distance per time t:

$$Velocity = distance \,/\, time$$

So velocity is described in corresponding units such as feet per second (fps). For example, the distance could be the length of flow down a drainage system; for a given velocity, the time of travel can be found by solving the preceding equation for time. Or the velocity could be the velocity of a ponded water surface approaching the soil as the water infiltrates; to limit the ponding to a given length of time, solve the equation for distance to find the depth to which water can be ponded.

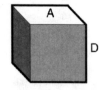

Figure 3.2   Volume = AD.

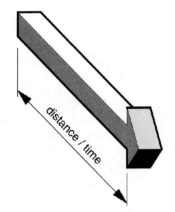

Figure 3.3   Velocity is distance per time.

Analytical Exercises

1. A common velocity of water in urban streams is 4 feet per second. Convert this to miles per hour. Also estimate your typical walking speed, and express it in miles per hour.

2. Find the length of a typical urban stream in your area, or that of a stream you know well, from headwaters to mouth. If water in the stream flows typically at 4 feet per second, how long does it take to flow the entire length of the stream and discharge from the watershed?

FLOW

Rate of flow is derived from combining the concepts of volume and velocity. If you dump a bucket of water on the road, you could say that you have created a flow of one cubic foot. If you keep on dumping more buckets, all the volumes of water flowing down the gutter add up to a stream of volumes over time (Figure 3.4). You can express the

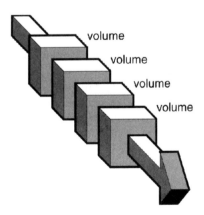

Figure 3.4   Rate of flow, q, is a stream of volumes over time.

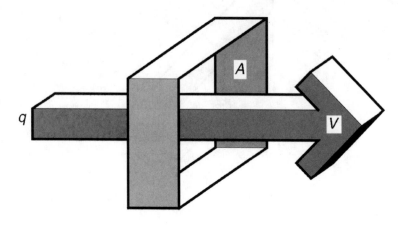

Figure 3.5 $q = VA$.

stream's magnitude in units such as cubic feet per second (cfs). Any culvert, any swale, any river, any water supply pipe has a flow that you can express the same way. A given volume of water flowing in a given amount of time produces the rate of flow q:

$$q = \text{volume} / \text{time}$$

Because volume = area × depth, and velocity V = distance or depth per time, we can also say (Figure 3.5),

$$q = \text{area} \times \text{velocity} = AV$$

For example, the area A might be the cross-sectional area of a pipe. If the velocity V is not allowed to exceed a certain limit, then the allowable rate of flow q is area times allowable velocity. Or if water is soaking into the earth at a downward velocity V equal to the soil's permeability and a given rate of flow q must be maintained, then the area of soil the water needs for soaking in can be derived by solving the preceding equation for A. Typical units in this equation would be fps for velocity, cfs for flow, and square feet (sf) for area.

One of the ways flow is produced is by pressure (Figure 3.6). "Head," or depth of water, produces pressure that pushes water through an orifice, over a weir, or even across a level plane. Use of the term *head* for hydraulic pressure may have originated in old water mills, where there was headwater above the mill and tailwater below and it was convenient to describe available hydraulic pressure in terms of feet of head coming down the pipe.

Total volume of flow results when a rate of flow has gone on for a period of time. By simple algebra, volume of flow Q_{vol} is the product of the rate of flow (volume/time) and the time t over which the flow continues (Figure 3.7):

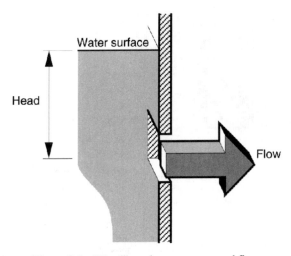

Figure 3.6   "Head" produces pressure and flow.

$$Q_{vol} = qt$$

Volume of flow is expressed in cubic units, such as cubic feet or acre-feet (af), because you have arrived again at a static volume.

Analytical Exercise

1. Using algebra and basic concepts of volume and time, derive the equation $q = AV$ from the equation $q = $ volume/time.

STORAGE

The change in volume stored in a body of water, such as a stream, a reservoir, a subsurface aquifer, or your bathtub, is the difference between inflow and outflow volumes over a

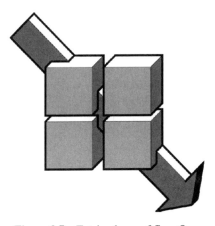

Figure 3.7   Total volume of flow Q_{vol}.

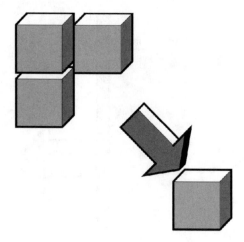

Figure 3.8 Change in volume of a body of water = $Q_{volin} - Q_{volout}$.

period of time. Like other volumes, it is expressed in volumetric units such as acre-feet. If Δ, the Greek letter delta, signifies difference or change, then (Figure 3.8),

$$\Delta\text{Storage} = Q_{volin} - Q_{volout}$$

The preceding equation shows why the water level in a stormwater reservoir rises when the rate of inflow is high, and falls when the inflow declines after a storm is over. It also shows why the amount of water stored in a subsurface aquifer declines when impervious cover reduces percolation of water into the soil, while the outflow continues at its natural rate.

If the volume of water that leaves a reservoir following a storm is equal to the volume that flowed in while the rain was falling, then,

$$\Delta\text{Storage} = Q_{volin} - Q_{volout} = 0$$

In this type of reservoir, the storage of water is only temporary. Temporary storage does not make water disappear. It only changes how fast a given volume Q_{vol} is moving around. The total volume eventually makes its way downstream.

Each unit of volume resides in a stream, pool, or reservoir for a certain time between entering the body of water and, later, leaving it. Residence time t_r is expressed in time units, such as hours or days. When the rates of inflow and outflow are equal, residence time is the ratio of pool volume and rate of throughflow:

$$t_r = \text{pool volume} / q$$

So the length of time water has to settle out its pollutants in the still water of a pool or wetland is equal to the volume of water in the pool, divided by the rate of throughflow. Equally, the amount of time water has to infiltrate and cleanse itself in contact with soil and vegetation as it flows down a swale is equal to the volume of water in the swale, divided by the rate of flow.

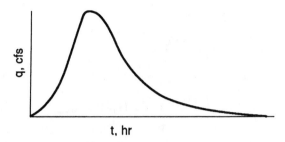

Figure 3.9 A hydrograph is a plot or table of q over time.

Analytical Exercise

Using algebra, derive from the equation t_r = pool volume / q an equation that expresses t_r as a function of, among other things, the cross-sectional area through which the water flows. Simplify the equation as much as possible.

HYDROGRAPH

A hydrograph is a useful summary of stormwater flows. It is a plot or table of rate of flow q over time t. Figure 3.9 shows a typical hydrograph during a storm.

At the beginning of a storm runoff q is very low. As the storm progresses, runoff builds to a peak. As shown in Figure 3.10, the rate of flow at the peak moment is designated q_p; the time when it occurs is t_p. As the storm recedes, the rate of flow falls to its original low level. The overall shape of the curve is typically not "bell shaped"; the peak occurs relatively early in the event, and the recession after the peak is relatively long.

A hydrograph also implicitly shows Q_{vol}. Imagine that you put the hydrograph through a bread slicer, cutting it into many thin vertical slices like those shown in Figure 3.11. Each slice approximates a tall, skinny rectangle, and each is a few minutes in width. The rate of flow during that moment is shown by the height of the curve at that slice. The Q_{vol} in each slice is q multiplied by time, which is the area of the rectangular slice (width × height). Q_{vol} for the storm as a whole is the sum of the Q_{vol}s in all the slices, which is the area under the whole runoff curve: q extended over a period of time, like that shown in Figure 3.12.

The shape of a typical storm hydrograph is close to that of a triangle, like that shown in Figure 3.13. Treating a hydrograph this way allows the quantities shown in the hydrograph

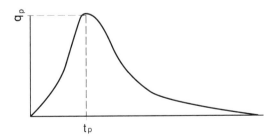

Figure 3.10 A hydrograph shows peak flow and time to peak.

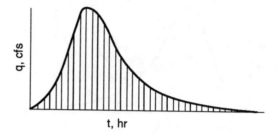

Figure 3.11   A hydrograph divided into slices for analysis of Q_{vol}.

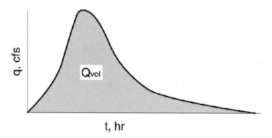

Figure 3.12   A hydrograph shows total volume of flow Q_{vol}.

to be analyzed with the simple geometry of the triangle. The area to the left of t_p is a right triangle; the area to the right of t_p is another right triangle. Thus, Q_{vol} can be found as the sum of the areas of the two right triangles.

Analytical Exercises

1. Figure 3.14 shows the triangular hydrograph shape assumed by the U.S. Soil Conservation Service (1972, pp. 16.5–16.6). Peak flow q_p occurs at time t_p; the time of receding flow following t_p is equal to $1.67 \, t_p$. Using the geometry of right triangles, derive an equation for the total volume of flow Q_{vol} in this hydrograph, given q_p and t_p.

2. In the same hydrograph, what proportion of Q_{vol} occurs during the rising flow before (to the left of) t_p? What proportion occurs during the receding flow after (to the right of) t_p?

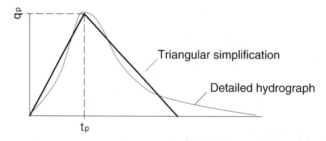

Figure 3.13   A triangular shape simplifies hydrograph analysis.

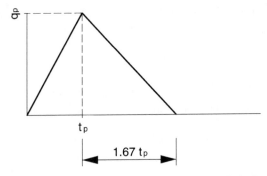

Figure 3.14   Triangular storm hydrograph used in analytical exercise.

3. In the same hydrograph, what proportion of Q_{vol} occurs at rates of flow higher than $2/3\ q_p$? What proportion occurs at rates higher than $1/3\ q_p$?
4. If Q_{vol} after development is greater than that before development, but time to peak t_p does not change at all, in what way does q_p change? Write an equation describing q_p as a function of Q_{vol} and t_p.

CONSTITUENTS

The concentration of a constituent in water is the ratio of the amount of the constituent to the amount of water (Figure 3.15). A common unit is weight of constituent per volume of water, milligrams per liter (mg/l). Because the metric system links the volume and mass of water, one mg/l at ordinary temperature is equal to one part per million (ppm, 0.1 percent) by weight.

The total mass of a constituent (mg) is equal to the concentration (mg/l) times the volume of water (l):

$$\text{Mass} = \text{concentration} \times \text{volume}$$

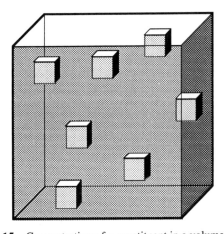

Figure 3.15   Concentration of a constituent in a volume of water.

The flow rate of a constituent (mg/day) is equal to the amount of constituent (mg) divided by the period of time over which the flow occurs (days):

$$\text{Flow rate} = \text{mass} / \text{time}$$

Combining some of the preceding concepts, flow rate of a constituent (mg/day) is also equal to the concentration (mg/l) times the rate of flow of the water that carries it (l/day):

$$\text{Flow rate of constituent} = \text{concentration} \times q$$

Analytical Exercise

Write an equation for flow rate of a constituent in lb/day as a function of concentration in ppm and water's flow in cfs. Use the equivalencies in Table 3.1 to convert from metric units.

MANAGEMENT ALTERNATIVES

With the background gained so far, you are ready to start considering stormwater management alternatives and doing the calculations to design for them on specific sites.

Estimating hydrologic flows is the place to start. Flows can be estimated for short-term peak storm events or long-term average flows. Select types of hydrologic analysis to inform the types of flow processes that you want to control through design. Estimation is done for specific selected locations on a site, under explicit assumptions of the conditions that may occur.

The alternatives for management are functionally different scenarios according to how water moves through a site and through the environment. You have to design specifically for the types of effects you want to bring about. Each management alternative can implement a unique combination of objectives for stormwater control, conservation or restoration. Many site developments end up with a mixture of processes, applied to appropriate places in the drainage system.

Conveyance

Conveyance (Figure 3.16) is the moving of surface runoff from one place to another. It ends with discharge to off-site streams, lakes, or bays.

The facilities for conveyance are pipes and channels. Even pipes that are buried under the ground are part of surface conveyance systems, because their impervious sides and bottoms prevent water from infiltrating the surrounding soil; they allow it to move only laterally, as all surface water does.

Conveyance is an ancient practice. Among the ruins of the ancient Roman city of Pompeii you can see streets with gutters that drain continuously to discharge points.

In the modern industrial world, Frederick Law Olmsted was a pioneer of stormwater conveyance when, in 1869, he implemented the process in the new community of Riverside, Illinois. In those days the streets where people walked were poorly drained and often covered with mud and manure. Olmsted stated that this was a nuisance problem, an aesthetic problem, a public health problem. He said the better alternative was to drop the stormwater

Figure 3.16   Conveyance.

and all the filth it carried off the streets, into a system of buried pipes. If you go to River-side, you can still see the stormwater inlets he installed in the streets and trace the pipes to their discharge in the Des Plaines River. Sanitation was one of the great humanitarian efforts of modern civilization. Conveyance solved the problem of Olmsted's time.

More than a century later, unmitigated conveyance no longer has an exclusive place in responsible site development. In today's world we know more about the effects of imper-vious surfaces and our way of life on water quality, groundwater, and rivers, and we are just as concerned about what happens to the filth in the Des Plaines River as we are about the health of residents in our own communities. These concerns have caused us to consider alternative scenarios for stormwater management.

However, conveyance is still an essential tool for managing the flows and processes of water. It controls day-to-day throughflows and their contact with soil and vegetation. For every drainage facility, of every type, an overflow conveyance system must be in place to discharge the excess flows from large, rare storms while preventing them from causing on-site flood and erosion damage. Therefore the matter of conveyance will keep coming up again and again, no matter what your ultimate hydrologic objective is.

Detention

Detention (Figure 3.17) is the slowing down of surface flows as they move away. The basic facility is a storage reservoir with a constricted outlet. Its purpose is to suppress downstream flooding and erosion by reducing the rate of flow. Although it reduces the peak flow rate q_p of runoff at the point of discharge, the total volume of flow Q_{vol} is still allowed to run downstream, stretched out over time.

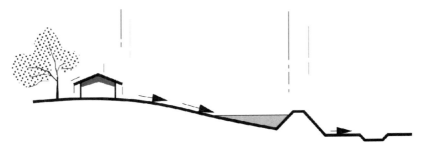

Figure 3.17   Detention.

Detention has been widely practiced in metropolitan areas of the United States since about 1970. It responds to the increase in peak rate of storm flow that was found, in the 1960s, to accompany most urban development. Although the application of detention to urban development is only a few decades old, it builds, to a degree, on experience in regional flood control that goes back to the beginning of the twentieth century.

Detention is a relative, quantitative modification of conveyance, because it is still qualitatively a discharging of runoff at the surface. Detention is capable of suppressing flood peaks when it is properly applied on a watershed-wide basis, but when it is applied indiscriminately its results have been discouraging. Detention, alone, is inherently unable to address water quality, groundwater replenishment, or water supplies. It will have a place in much site development for the foreseeable future. But it cannot be considered a fundamental solution of the environmental problems that we know urban development presents. Other alternatives must be considered.

Extended Detention

For water quality control, extended detention has become popular. When water sits still in a pond or wetland, suspended particles can settle out and chemicals can be adsorbed (adhered onto particle surfaces) in bottom sediment, taken up by biota, and biodegraded. A pond for treatment may have to be larger than one needed for flood control alone, so as to give constituents adequate residence time for the treatment processes to be completed.

A pond or wetland is a land-based "treatment plant." Land-based systems take advantage of the free natural filtering and transforming capacities of light, air, soils, and organisms. Although they require land in the midst of urban developments, they do not require a lot of expensive artificial materials or maintenance as compared with mechanical treatment plants.

Like storm detention, extended detention is a modification of conveyance. It eventually discharges all runoff to off-site streams. It is capable of improving surface water quality when it is specifically and knowledgeably designed for this purpose. However, its inability to adequately address volume of runoff, groundwater, or water conservation leads us to consider still other alternatives.

Infiltration

Infiltration is the soaking of water into the ground (Figure 3.18). The water no longer moves laterally across the surface. Infiltration is qualitatively different from conveyance and

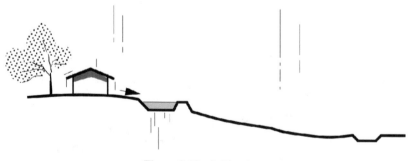

Figure 3.18 Infiltration.

detention because it makes water go to a different part of the environment, where it under-goes different types of processes. It addresses flooding and erosion as well as water quality, groundwater, and water supplies. This is inherently the most complete solution to stormwa-ter issues in the environment, because it restores natural hydrologic processes.

Infiltration happens, to a degree, in all vegetated swales and porous soil surfaces. It hap-pens particularly in an infiltration basin, an earthen basin with no primary surface outlet. An infiltration basin can be built either on its own or in the base course of a pavement.

Experience with infiltration basins has been gained for more than a half century on Long Island, where it is known by the name of "recharge," because it replenishes underly-ing groundwater. In other parts of the country its practice has fluctuated as various programs have been advocated and experiments have been tried. With a wealth of research and expe-rience now to support is use, infiltration is becoming deservedly more common.

Infiltration almost always improves water quality. Almost all soils filter out suspended particles and adsorb chemicals. The ecological processes in soils degrade pollutants. Almost all of the filtering and transformation processes take place in the top few inches of soil, pre-venting constituents from contaminating underlying groundwater. Like an extended-deten-tion pond, the soil where water infiltrates is a kind of treatment plant.

Water Harvesting

Water harvesting is the direct capturing and using of runoff on-site (Figure 3.19). In some applications, water harvesting maintains the water levels in permanent ponds and wet-lands. During dry months, just enough may be supplied to balance a pond's water level. In wetter months, surplus water passes on through the pond's outlet.

In other applications, water harvesting supplies water for irrigation. A simple method is to let runoff soak into the soil in infiltration basins where plants are rooted. A more elabo-rate way is to store harvested runoff for days or weeks in tanks or reservoirs, then release it into irrigation lines when the soil is dry.

Water harvesting is an ancient practice. It can be seen on the rooftops and in the urban plazas of cities of classical and medieval civilizations. For centuries, Native Americans in the arid Southwest have diverted ephemeral streams to their orchards and fields.

In the modern industrial world this scenario has only recently been rediscovered and adapted to our new way of life. Ponds can now be analyzed before construction to find out whether an adequate runoff supply is, in fact, present. In Arizona and Colorado water har-vesting has supported the irrigation of urban landscapes for facilities ranging from branch

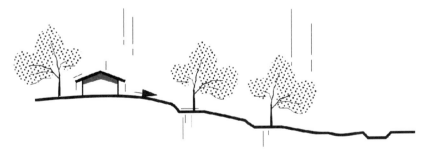

Figure 3.19 Water harvesting.

banks to new communities. As our demands for water become greater while we become more sensitive to the sustainability of development, water harvesting may become an important part of our future.

COMMENTARY

The chapters on applications (Chapters 6 through 10) are organized according to the types of hydrologic processes that the various management alternatives encourage. Each alternative can be designed to meet the needs of specific sites and the people who live there.

In describing management alternatives, this book avoids the use of the term *retention*, because this term has been used inconsistently throughout the country. In various times and places I have seen it used to refer to detention, extended detention (with or without a permanent pool), a permanent pool (with or without any particular rate or quality effect), infiltration, and two or more of these approaches at once. When a word has been used to mean mutually exclusive things, it ends up having no consistent meaning at all. Its uselessness is multiplied when even the people who are using it do not know what it means. I once asked a designer in Florida about the function of the "retention" basins he had frequently designed under ordinances requiring basins of certain capacity. "Water does not just go into a basin and disappear," I urged. "Where does the water go; what physical process in the water does the basin bring about?" "We don't know what happened to the water," he responded. "We retained it." So I do not use the word *retention* in my own work. If in your practice you work under a local ordinance that uses the term *retention* with a single, clear meaning, then you can use the term that way in that locale, but do not expect people elsewhere to understand the word the same way. The chapters in this book on management use other functional terms, such as *conveyance*, *detention*, and *infiltration*; each defines, exclusively and unambiguously, a single specific physical process.

REFERENCES

Debo, Thomas N., and Andrew J. Reese, 1994, *Municipal Stormwater Management*, Boca Raton: Lewis Publishers.

Leopold, Luna B., 1974, *Water, A Primer*, San Francisco: Freeman.

U.S. Soil Conservation Service, 1972, *National Engineering Handbook,* Section 4, Hydrology, SCS/ENG/NEH-4, Washington: U.S. Soil Conservation Service.

Van der Leeden, Frits, Fred L. Troise and David Keith Todd, 1990, *The Water Encyclopedia*, second edition, Boca Raton: Lewis Publishers.

CHAPTER 4

STORM RUNOFF

Storm runoff refers to the volumes and rates of flow in individual storm events. It is often called "direct runoff," because it results from surface flow and other immediate responses to precipitation.

Before you can design to control storm runoff, you must establish its magnitude. Every swale, pipe, basin, and pool must be designed for the runoff from its specific drainage area. You often have to estimate storm runoff both before and after development in order to deal appropriately with the impacts of land use change.

A runoff gauging station would provide a direct, factual way to observe flows from a site in its existing condition. But few development sites have gauging stations. Even if a gauging station were present, it would not by itself predict the flows that would occur after a proposed development is constructed.

Therefore, some sort of estimate is necessary, based on data about the site and general knowledge of runoff processes. The estimation of storm runoff is a modeling of a natural process.

Rainfall runoff models estimate runoff based on the drainage area that outlets at a point, and the rainfall upon it (Figure 4.1). In principle such models are valid, because quantity of runoff is known to vary with the quantity of rainfall and the size and condition of the watershed. In practice they are convenient, because there are many weather stations around the country that monitor rainfall, and it is presumed that rainfall records from a given station are valid for all sites in the surrounding locale.

THE DESIGN STORM

A design storm is a particular combination of rainfall conditions for which you estimate runoff and design a drainage system. The magnitude of a design storm may be expressed as

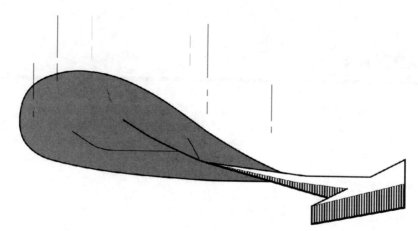

Figure 4.1 Concept of a rainfall-runoff model: from an inflow of precipitation, the drainage area produces an outflow of runoff.

a total quantity of precipitation such as inches of rainfall, or as a short-term intensity such as inches per hour.

Recurrence interval is a means of expressing the probability that a storm of a given size or intensity may occur at your site (Figure 4.2). Recurrence interval, or frequency, is the average time between storms of a given magnitude. A 10-year storm is large enough that it has recurred, on the average, in only one of every 10 years in the local rainfall record. A 100-year storm is so big that it has occurred only once every 100 years. For any given recurrence interval, the magnitude of the storm varies from region to region with the climate, as illustrated by the four curves in Figure 4.2.

The probability of occurrence in any one year is the reciprocal of the recurrence interval (Figure 4.3). The 10 year storm has a 10 percent chance of occurring in any one year; the 100-year storm has a 1 percent chance.

There is always a risk that, at any time, the design storm you select could be exceeded and stormwater facilities you design could be overloaded. Selection of a large, infrequent design storm reduces the risk, because it results in large drainage facilities that contain the large amount of water. But large facilities are expensive to construct. You must select the balance of risk and cost that is appropriate to each specific part of a drainage system. For culverts and detention basins that drain local streets and prevent local drainage problems, a 2- 25-year recurrence interval is common. For water quality and groundwater replenishment, tremendous good can be done in designing for storms even smaller than the 1-year storm—and because design for such small storms is quite feasible, they are frequently used for this purpose. On the other hand, in extremely sensitive situations, where people's homes would be seriously damaged by flooding or their lives would be endangered, the most appropriate storm is the maximum probable storm, which can be much larger than the 100-year storm. Local ordinances frequently govern the recurrence interval to be used.

On every site, two drainage systems exist side by side, the primary system and the secondary. The primary system operates during all storms up to and including the design storm. The secondary system begins to operate whenever a larger storm occurs or the primary system is clogged. In every drainage facility you will ever design in your life, in selecting a design storm you are accepting the fact that your pipe or swale or basin will overflow when a larger storm occurs or the primary facility is clogged.

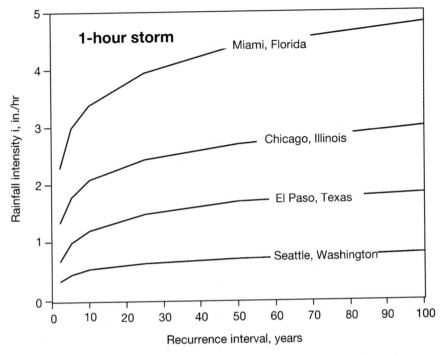

Figure 4.2 Intensity of one-hour storm increases with recurrence interval.

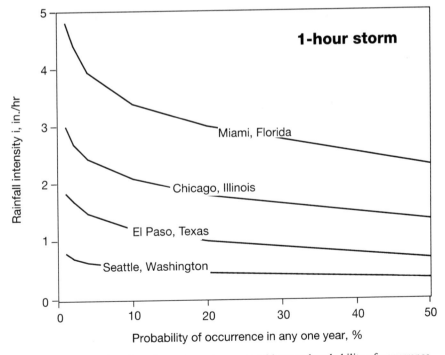

Figure 4.3 Intensity of one-hour storm decreases with annual probability of occurrence.

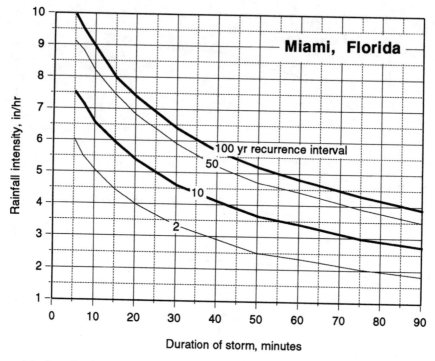

Figure 4.4   Intensity-duration-frequency curves for Miami, Florida (data from U.S. Weather Bureau, 1955).

Storms have durations as well as recurrence intervals. A storm that is short in duration can be very intense. As duration continues, high intensity cannot be maintained; the average intensity decreases. In specifying a design storm, you must specify both recurrence interval and duration.

Figures 4.4 and 4.5 are examples of graphs showing rainfall intensity as a function of both duration and frequency. The curves on this kind of graph are referred to as intensity-duration-frequency curves, or IDF curves. The scale at the bottom is duration. You can read up from a selected duration to the line representing recurrence interval, thence horizontally to the left to find the intensity corresponding to that combination of duration and frequency. There is a unique set of IDF curves for each weather station.

Analytical Exercises

1. What is the probability that the 2-year storm will occur in any one given year? The 50-year storm?

2. This extended exercise exposes you firsthand to the scientific data on which the precipitation characteristics of design storms are based and sensitizes you to the data's variability. In a library, find the daily weather record for a weather station in your locale. In the United States, this kind of data is published by the National Oceanic and Atmospheric Administration (NOAA). Find the largest single daily rainfall in each of the most recent 10 years. Rank these by size, with 1 being the largest and 10

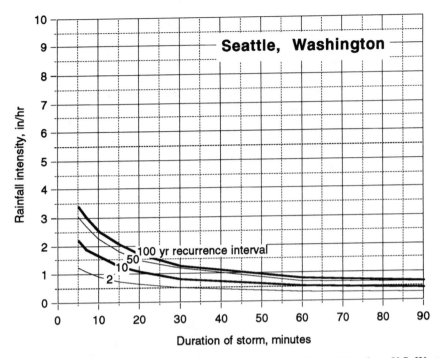

Figure 4.5 Intensity-duration-frequency curves for Seattle, Washington (data from U.S. Weather Bureau, 1955).

the smallest. Find the recurrence interval RI of each one by using the formula $RI = N/M$, where N is the number of years in your record (10), and M is the rank of the individual item. Then do the same thing for the preceding 10-year period. How different are the two values for the 2-year storm? How different are the values for the 10-year storm? Which value is more "right"? Which one would you use in design? Is the 10-year result more variable than the 2-year result? Why? If you keep looking at successively older preceding 10-year periods, what kinds of values do you think you will find? How do you suppose the U.S. Weather Bureau came up with the curves represented in Figures 4.4 and 4.5? Be specific. How "right" are these curves?

THE DRAINAGE AREA

The drainage area, or watershed, is the land area that drains to the point at which you estimate runoff. Any rainfall runoff model requires you to identify the drainage area and to specify its size, soil, and condition.

Drainage Area Boundary

A drainage area is identified by defining its boundaries on a map. Do not confuse a drainage area's boundaries with prominent ridge lines; they may or may not coincide. The boundaries are identified by systematically applying very specific criteria.

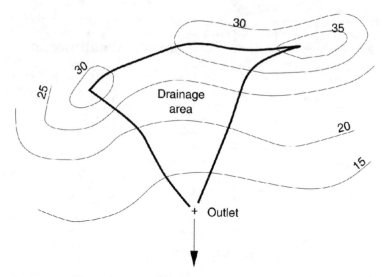

Figure 4.6 A drainage area is defined by its outlet.

A drainage area is defined by the point that it drains to (Figure 4.6). Therefore, before doing anything else, identify exactly the location where you are going to estimate runoff. The watershed you wish to identify is the one that drains to this point. On one side of this watershed's boundary, a running drop of water would flow eventually through the outlet—the area on that side of the line is inside the drainage area. On the other side of the line a running drop of water would miss the outlet and drain elsewhere—that side is outside the drainage area.

The topographic contours uphill from the outlet tell you where the drainage area's boundary is. Before development, follow existing contours. After development, follow proposed contours.

On most sites, the contours near the outlet indicate two sides of the drainage area, diverging uphill from the outlet point (Figure 4.7). Draw the boundary starting exactly

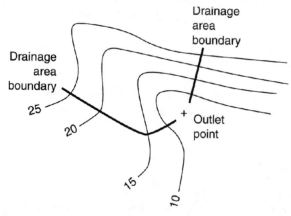

Figure 4.7 Two sides of a drainage area's boundary near the outlet point.

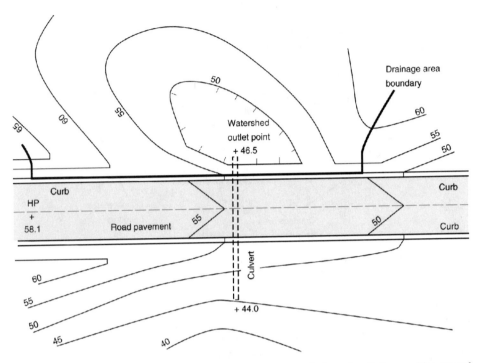

Figure 4.8   Where the watershed outlet is a topographic sump, the boundary follows a ridge around the outlet.

from the outlet and moving uphill on each side by the shortest (steepest) possible path. Continue via the shortest path from each contour to the next. Gently curve the boundary line so that it crosses each contour at a right angle.

On some other sites, the outlet of the drainage area is a topographic low point or sump, such as an infiltration basin or the mouth of a culvert like that in Figure 4.8. On these sites the lower boundary of the drainage area is a ridge around the outlet. Find the part of that ridge nearest to the outlet and use it as the beginning of the boundary's identification. From this point, draw the boundary uphill in each direction by the shortest (steepest) path, from contour to contour, crossing each contour at a right angle.

When you arrive at a topographic high point, such as the peak of a hill or the ridge of a roof, stop working on that part of the boundary for a minute.

Start again at the outlet and go uphill in the other available direction. Again, keep going until you reach a high point.

If you end up connecting with the part of the line that you previously drew, then you are done; your boundary wraps completely around the watershed. You should be able to step back from your map and perceive, as a three-dimensional form, the land form that is draining water toward the outlet.

If your lines do not connect yet, then your lines have reached two separate high points. You have to connect the high points through one or more saddle points. As shown in Figure 4.9, a saddle point is the low point along a ridge connecting two high points. On two sides of a saddle point the topography goes up toward the two high points. On two other sides the topography goes down the hillsides. You can imagine sitting in the "saddle" with your legs

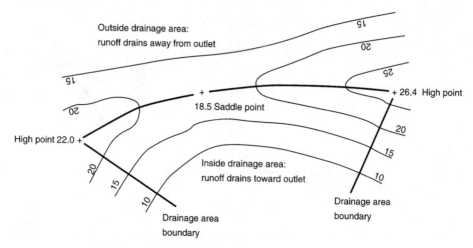

Figure 4.9   Drainage area boundary at a saddle point.

hanging down the two low sides, and the front and rear of the saddle rising up in front and in back of you.

Near one of your high points, identify a saddle point where water would flow downhill toward the drainage area's outlet on one side, and away from it on the other. This saddle is a local low point along the drainage area's boundary. Continue drawing your watershed boundary from the saddle point, working up the contours toward each high point, just as you originally did from the watershed outlet.

If necessary, continue a succession of lines, from each saddle point to each high point, until they connect and wrap completely around the area that drains to the outlet.

Structures like buildings and roads are special cases in the application of the preceding principles. You are likely to encounter plenty of them when analyzing a drainage area after development.

As you draw a watershed boundary line, working uphill from contour to contour, approach a building from the contours below. Follow the edge of the building up to its roof ridge line, as shown in Figure 4.10. The roof ridge is a high line in the building's "topography," and you analyze it as you would a topographic high point. When you reach it from one side, stop drawing for a minute and start considering how to approach the building from the other side. There will probably be a high point in a swale on the uphill side of the building. A swale's high point is a topographic saddle point. Start there and work up toward the building's ridge line, meeting the line you drew before. Then you can continue the drainage area's boundary by working uphill from the swale's high point in the opposite direction, as you would from any other saddle point.

Similarly, approach a road from the low side, following the contours uphill as usual, as shown in Figure 4.11. Where the road has a curb, the curb on the low side of the road is likely to be a local topographic ridge line. Go to the top of that curb and follow it uphill to the road's high point. If the road has no curb, the road's crown is probably the topographic high line instead—so go to the crown and follow it as you would a curb or a roof ridge line. If the topography continues uphill on the far side of the road, cross the road at its high point, then let the topographic contours continue to guide you up the hillside.

If you have any doubts about whether you have drawn part of a watershed boundary correctly, you can test it by imagining a drop of water just to one side of the boundary starting

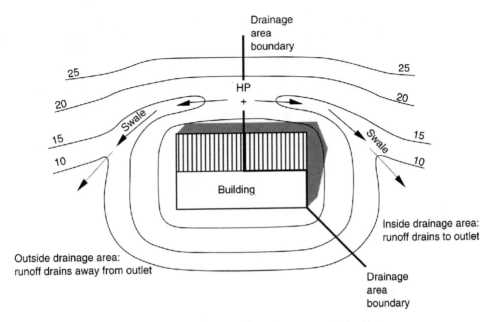

Figure 4.10   Drainage area boundary at a roof ridge line.

to flow overland. On one side of the boundary the drop should run ultimately to the watershed's outlet; on the other side it flows away somewhere else. That is the definition of a drainage area's boundary, and the test of whether you have drawn it correctly.

Size, Soil, and Cover

Having drawn a drainage area's boundary correctly, you are in a position to estimate its size and to characterize its land use and soils to meet the needs of the rainfall runoff model you are using.

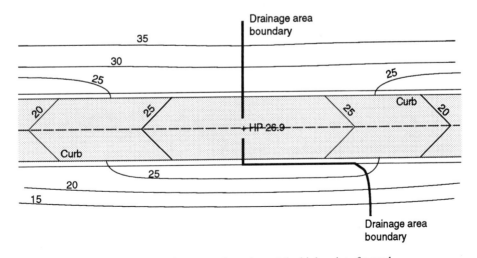

Figure 4.11   Drainage area boundary at the high point of a road.

TABLE 4.1   **Values of *c*, the cover factor or runoff coefficient in the Rational formula. For flat slopes or permeable soils use the lower values; for steep slopes or impermeable soils use the higher values (from Table 1 of *Design of Roadside Drainage Channels*, U.S. Federal Highway Administration, 1973; and Table 1 of *Airport Drainage*, U.S. Federal Aviation Agency, 1965.)**

Type of Surface or Land Use	c
Individual soil covers	
Forest	0.1–0.3
Turf or meadow	0.1–0.4
Cultivated field	0.2–0.4
Steep grassed area (2:1)	0.5–0.7
Bare earth	0.2–0.9
Gravel or macadam pavement	0.35–0.7
Concrete or asphalt pavement	0.8–0.9
Composite land uses	
Flat residential, about 30% impervious	0.40
Flat residential, about 60% impervious	0.55
Sloping residential, about 50% impervious	0.65
Sloping, built-up, about 70% impervious	0.80
Flat commercial, about 90% impervious	0.80

The size in acres can be scaled from the site map. The area of impervious surfaces can be similarly scaled, then expressed as a percentage of the total area.

Tables 4.1 and 4.4 (see pages 67–68) list factors representing soil and cover conditions in rainfall runoff models that are discussed in detail in the following sections. If your watershed is homogeneous with only one soil and land use condition throughout, then you can just look up the factor for that condition in the appropriate table. The tables also list values for certain composite land uses. Where conditions in your watershed closely resemble one of those described in the relevant table, you can read or interpolate the table's listings directly.

Where there are two or more very different soil types or land uses as in Figure 4.12, you have to derive a composite factor representing the watershed as a whole. First measure the area occupied by each land use or soil type, and read its numerical factor from Table 4.1 or 4.4. Then find the average factor for the drainage area, weighting the average according to how much of the total area each factor occupies. If the factors for three soils are A, B, and C, then the weighted average factor for the three soil areas is given by:

$$\text{Weighted average factor} = \frac{A(\text{area of A}) + B(\text{area of B}) + C(\text{area of C})}{\text{total drainage area}}$$

The resulting average applies to the drainage area as a whole. You can use the same type of weighted averaging for any number of soil and land use conditions.

When comparing runoff before and after development, you should understand that almost any characteristic of a drainage area can be altered by development, even though the watershed's outlet remains at the same place. Changes in land use always involve change in impervious cover and vegetation, and in the resulting runoff coefficients and speed of runoff travel.

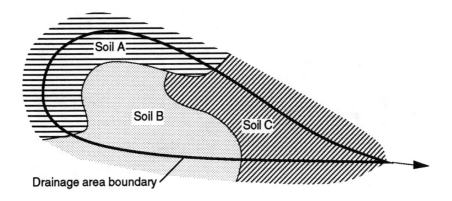

Figure 4.12   Different soils in a drainage area.

Drainage area boundaries can change as a result of earthwork and the installation of roads and buildings. Roads have gutters, swales, or storm sewers that divert runoff across hill slopes. The roofs of large buildings can pitch runoff in significantly new directions. New boundaries mean that a drainage area's size and time of concentration may be altered. Soil type and slope may be altered if the new boundaries swing the drainage area into parts of the site with different land forms.

The descriptions of exercise sites in the Appendix show drainage areas' size, slope, and soil essentially unchanged by development. That is a plausible assumption for the medium-density residential or office type of development assumed in the developed land use. In practice, you should check every relevant condition, beginning with drawing drainage area boundaries, both before and after development.

Time of Concentration

Runoff's travel time is one of the watershed characteristics that can strongly influence the rate of storm flow. If a given volume of runoff drains off a drainage area quickly, then the peak rate of flow at the outlet is correspondingly high.

Generically, travel time can be expressed by water's velocity and the distance it has to travel:

$$t = L / V$$

where

t = travel time in minutes

L = length of travel in feet: can be scaled from a site map

V = velocity in feet per minute

High velocities tend to occur on smooth surfaces such as pavements; high velocities produce high rates of runoff. Low velocities tend to occur on rough, heavily vegetated surfaces; low velocities produce slow, gentle runoff. Velocities can change as runoff flows downslope, passing first through rough undergrowth, then concentrating into a smooth channel.

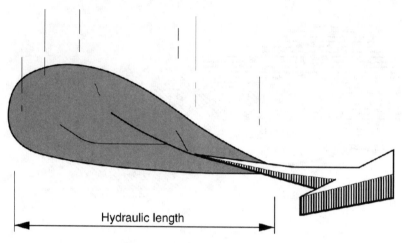

Figure 4.13 Hydraulic length.

Time of concentration is a special case of travel time. It is the maximum amount of time runoff from any point in a drainage area takes to flow to the outlet. Among a number of alternative paths that runoff could take from distant parts of a watershed, time of concentration is defined by the longest possible time, whether or not it involves the longest distance. The line along which that runoff travels is the drainage area's hydraulic length (Figure 4.13). Most storm runoff models use time of concentration to indicate how fast water could drain off a site and contribute to peak rate of flow. Each model includes its own appropriate procedure for estimating and using time of concentration.

Analytical Exercise

On a site map, identify the discharge point of a development site. Outline its drainage area. Using ordinary scaling methods, determine the area of the watershed and the percentage of impervious cover in the development.

THE RATIONAL METHOD

The Rational Method is about 100 years old (Kuichling, 1889). This rainfall runoff model came to the United States during the early development of New York State, where thoughtlessly installed storm sewers were frequently overloaded and a "rational" method of sizing culverts was called for. The method's simple equation did provide such a method for the first time, and the name stuck. The Rational Method is still widely used for stormwater design because it is simple to use. It is good for teaching the fundamentals of rainfall runoff models because it condenses a model's essential components into an easy-to-understand form. Because the Rational Method was developed to estimate peak flows from small urban drainage areas, its application today is usually restricted to drainage areas of less than 200 acres.

The Rational Method associates peak rate of runoff with three easily identifiable characteristics of a drainage area and the rainfall upon it. The variables are:

A_d = drainage area (acres); this can be scaled from a site map

c = the drainage area's cover factor or runoff coefficient (no units), based on a combination of soil, land use, and slope; values for c are given in Table 4.1

i = rainfall intensity (in./hr) at a selected recurrence interval and duration

A simple formula connects the three variables. With q_p being peak rate of runoff in cfs,

$$q_p = A_d \, ci$$

Keep the units straight: multiplying the units in the three variables does not automatically give cfs. In the early days there was an additional factor for correcting the units, equal to 1.008 cfs per ac-in./hr. Because this factor made no appreciable difference to the result, everyone just forgot about writing it into the formula. Now we are left with a very short, simple formula, but one in which you have to make sure you use the correct units for each variable.

For obtaining the value of i, rainfall intensity, IDF curves for selected locations are shown in Figures 4.4 and 4.5 and in the Appendix. Similar graphs for about 200 locations are presented in the U.S. Weather Bureau's Technical Paper No. 25, *Rainfall Intensity-Duration-Frequency Curves* (1955) and in local sources such as the 1987 article by Aron and others for Pennsylvania cited at the end of this chapter. For sites without specific rainfall charts, you can interpolate rainfall geographically from charts of nearby cities as long as there is no intervening feature that may alter rainfall, such as a mountain, a large lake, or a substantial change in elevation.

The Rational formula assumes that, for a given recurrence interval, a storm duration that matches a drainage area's time of concentration produces the greatest rate of runoff. The idea is that a shorter duration would indicate a more intense storm, but the rain would stop before runoff from all of the watershed arrived at the outlet to contribute to peak discharge. A storm with longer duration would continue adding rainfall as long as the runoff kept traveling to the outlet, but the rain would be less intense. A duration exactly equal to the time of concentration indicates the most intense storm that keeps contributing rainfall for as long as it takes for the watershed to concentrate runoff on the outlet.

Therefore, in selecting a value for i, time of concentration t_c must first be estimated by an appropriate procedure.

The Federal Aviation Agency (FAA) devised a relatively simple t_c estimation method for use with the Rational formula (U.S. Federal Aviation Agency, 1965). The FAA method finds time of concentration where overland flow occurs over any uniform slope and surface type by the following equation:

$$t_c = [1.8 \, (1.1 - c) \, L_h^{\frac{1}{2}}] / G^{\frac{1}{3}}$$

where

t_c = time of concentration in minutes

c = cover factor in the Rational formula; values are listed in Table 4.1

L_h = hydraulic length in ft

G = slope along the hydraulic length, in percentage

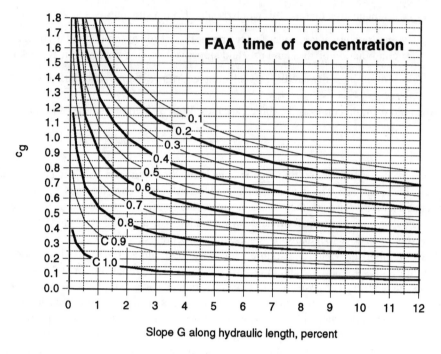

Figure 4.14   The factor c_g in the FAA method of estimating time of concentration (U.S. Federal Aviation Agency, 1965). Time of concentration $t_c = L_h^{1/2} c_g$, where hydraulic length L_h is in feet.

In order to get FAA's equation into a usable chart, I rewrote it as $t = L_h^{1/2} c_g$, where the factor c_g combines everything in the equation but $L_h^{1/2}$. The chart in Figure 4.14 shows the result. To use the chart, enter from the bottom with the slope, read up to the line for the watershed's cover factor c, thence to the left to read c_g. Then apply the factor to find overland flow time of concentration, $t_c = L_h^{1/2} c_g$.

If the hydraulic length passes over surfaces with substantially different slopes or values of c, you must determine the travel time for each homogenous portion of the hydraulic length separately, using the FAA method. Then the total time of concentration is the sum of the times on the various surfaces.

When the Rational Method was originally developed, it was intended only to identify the peak rates of flow that pipes and culverts had to carry. It translates peak intensity of rainfall directly into peak intensity of runoff. The method was not originally intended to estimate total volume of flow during a storm, although some designers have added supplemental assumptions to "stretch" the method to give volume. For applications that involve volume of flow, the SCS Method is preferable because it was originally developed to include it; the SCS Method is described in the next main section of this chapter.

A number of different versions of the Rational formula are being applied in practice. The variations have evolved over a century of the Rational Method's widespread use. In estimating time of concentration, some versions assume only overland flow time while others use combinations of overland and channel flow conditions. Several different tables of runoff coefficients can be found, giving slightly different values of c for similar watershed conditions. Various practitioners can get different q_p estimates for the same watershed, depending on their assumptions. You can find many different personal judgments about which version is best. Variability in application can be considered a weakness of the Rational Method. Further details about the Rational formula can be found in references such as those listed at the end of this chapter.

Rational Method Exercises

Exercises 4.1 and 4.2 find peak rates of runoff before and after development, so the results can be compared. They assume that the watershed before development is homogeneous, and that after development it is a mosaic of pervious turf and impervious roofs and pavements.

Fill in the exercise blanks with your data and calculations for a site you are working on for practice, a site you are familiar with in your region, or one of the sites described in the appendix. Filling in the columns for both Sites 1 and 2 can generate enlightening discussions about the conditions that create stormwater and the constraints on its management. The results from Exercises 4.1 and 4.2 are used for design in Chapter 6.

Summary of Process

1. Estimate watershed area A_d in acres, from a site map.
2. Obtain cover factor c from Table 4.1. If there is more than one c value in the drainage area, then find the weighted average, weighting each c value by the proportion of the drainage area it covers.

Exercise 4.1   Rational Method, Before Development

	Site 1	Site 2
	Determining data	
Design storm recurrence interval (from local standards)	= _____ yr	_____ yr
Drainage area A_d (from site map)	= _____ ac	_____ ac
	Peak rate of flow	
Cover factor c (from Table 4.1)	= _____	_____
Slope G along hydraulic length (from site map)	= _____ %	_____ %
Time of concentration factor c_g (from Figure 4.14)	= _____	_____
Hydraulic length L_h (from site map)	= _____ ft	_____ ft
Time of concentration t_c $= c_g L_h^{1/2}$	= _____ min	_____ min
Rainfall intensity i at duration equal to t_c (from IDF chart)	= _____ in./hr	_____ in./hr
Peak rate of flow q_p before $= A_d c i$	= _____ cfs	_____ cfs

Exercise 4.2   Rational Method, After Development

	Site 1	Site 2
Determining data		
Design storm recurrence interval (from local standards)	= _____ yr	_____ yr
Determining data		
Drainage area A_d (from site map)	= _____ ac	_____ ac
Peak rate of flow		
Cover factor c for turf (from Table 4.1)	= _____	_____
Cover factor c for pavement (from Table 4.1)	= _____	_____
Composite cover factor c (by weighted average)	= _____	_____
Portion of hydraulic length L_h in turf (from site map)	= _____ ft	_____ ft
c_g factor for turf (from Figure 4.14)	= _____	_____
Time of travel on turf t_1 $= c_g L_h^{1/2}$	= _____ min	_____ min
Portion of hydraulic length L_h in pavement (from site map)	= _____ ft	_____ ft
c_g factor for pavement (from Figure 4.14)	= _____	_____
Time of travel on pavement t_2 $= c_g L_h^{1/2}$	= _____ min	_____ min
Time of concentration t_c $= t_1 + t_c$	= _____ min	_____ min
Rainfall intensity i at duration equal to t_c (from IDF chart)	= _____ in./hr	_____ in./hr
Peak rate of flow q_p after $= A_d c i$	= _____ cfs	_____ cfs

3. Obtain hydraulic length L_h and its slope G from a site map. If the hydraulic length crosses surfaces with substantially different c values or slopes, find L_h, G, and c for each homogeneous portion of the hydraulic length.

4. Obtain the c_g factor from Figure 4.14. If there are different values of c or G along the hydraulic length, find c_g for each homogeneous portion.

5. Find time of concentration t_c in minutes using $t_c = c_g L_h^{1/2}$. If there is more than one c_g factor, find time of travel for each one; the total time of concentration is the sum of the travel times on the various surfaces.

6. Obtain rainfall intensity i in in./hr, at a selected recurrence interval and duration equal to t_c, from one of the IDF graphs in Figure 4.4, Figure 4.5, the Appendix, U.S. Weather Bureau Technical Paper No. 25, or local data.

7. Compute peak rate of runoff q_p in cfs from $q_p = A_d\, ci$.

Discussion of Results

1. Which site produces the higher rate of runoff before development? Which produces the higher rate after development? What site characteristics contribute to this pattern of results?

2. What is the average velocity of runoff along the hydraulic length on each site, before and after development? What factors contribute to the differences between the sites, and between the sites before and after development?

3. Calculate the runoff relationship between the sites, q_p for site 2 ÷ q_p for site 1, before and after development. Does development make the sites hydrologically more different, or more alike? What site and development characteristics led to this kind of change?

4. For each site, calculate the change in runoff with development, q_p after ÷ q_p before. Which site produces the greater change in runoff with development? What site characteristics produced this kind of change?

5. Rate of rainfall during a storm is expressed in inches per hour or inches per 24 hours, but runoff is expressed in cfs. The difference in units of measurement tends to prevent us from appreciating relative magnitudes of runoff and the rainfall that produces it. From the results of the Rational Method before and after development, convert your site's peak intensity of rainfall in inches per hour to cfs using the conversion factor of 1.008 cfs per ac-in/hr:

$$\text{Intensity of rainfall in cfs} = 1.008\, i\, A_d$$

Then answer these questions: What is the difference between rate of rainfall entering the watershed and runoff leaving it, before and after development? What types of physical processes in the watershed cause this difference?

6. What are the duration and recurrence interval of your design storm after development? Divide the duration by the recurrence interval to find the average proportion of total elapsed time occupied by the storm, in hr/hr. Express the result as a percentage of time. If you construct a channel to carry the runoff from your design storm, and then visit the site on a randomly selected day of a randomly selected year, would you expect to see the channel full of flowing water? What would you expect to see?

THE SCS METHOD

The SCS Method was developed and documented entirely by the U.S. Soil Conservation Service (SCS), beginning in the 1950s. In 1994 the name of the SCS organization was changed to the Natural Resources Conservation Service (NRCS). In coming years practitioners may refer to the method by the agency's new name. However, as of this writing, the agency has not published any new hydrologic manuals under its new name, so this book refers to the agency's method by the familiar old name of SCS.

Over the years SCS has published a number of handbooks explaining its method, gradually updating the charts and sequences of steps. Currently, the most prominent and useful of the various documents is Technical Release 55 (TR-55), *Urban Hydrology for Small Watersheds* (U.S. Soil Conservation Service, 1986). An additional source for underlying hydrologic theory is the *National Engineering Handbook*, Section 4 (NEH-4), Hydrology (U.S. Soil Conservation Service, 1972). The agency is continuing to conduct research relevant to its method.

Since its inception the SCS Method has grown in prominence, tending gradually to replace the Rational Method in practice. It is not necessarily more accurate than the Rational formula where it has not been specifically correlated with locally observed runoff data. But it is more directly useful than the Rational formula for applications requiring an estimate of runoff volume, and its thorough documentation usually allows it to be used without a lot of disagreement between designers and the municipal engineers who review the designers' work. The method can be applied to drainage areas much larger than those to which the Rational Method is limited.

The SCS Method's disadvantage has been that it is more complicated to learn and to apply than the Rational Method, especially for finding peak rate of flow. The use of appropriate software eliminates much of the difficulty of applying the method, but not the difficulty of initially learning it. This book provides charts for approximate manual application. Software implementing the method is available from SCS and from a number of commercial vendors.

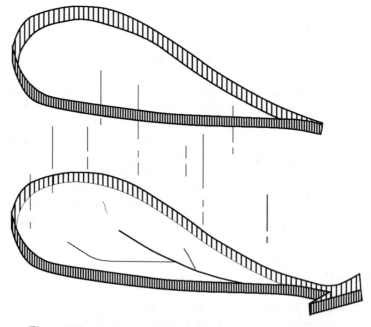

Figure 4.15 Depth of runoff produced by a depth of storm rainfall.

Runoff Volume

Unlike the Rational formula, the SCS Method begins by finding the total depth and volume of runoff during a storm. The depth of runoff is produced by the depth of rainfall (Figure 4.15). The rate of runoff in cfs is then determined by the distribution of rainfall intensity during the storm and how fast the given volume of runoff drains off the watershed.

The equation for runoff depth is the SCS Method's basic equation; it begins the entire method (Equation 2-1 of U.S. Soil Conservation Service, 1986):

$$Q_d = (P - I_a)^2 / (P - I_a + S)$$

where

Q_d = depth of runoff (in.)

P = depth of 24-hour rainfall (in.)

I_a = initial abstraction, the losses of rainfall to infiltration and surface depressions before runoff begins (in.); 0.2S is a more or less median value that has been found in the field

S = potential maximum retention after runoff begins (in.); it is defined by curve number CN, which is a function of the drainage area's soil and land use: $S = (1,000/CN) - 10$

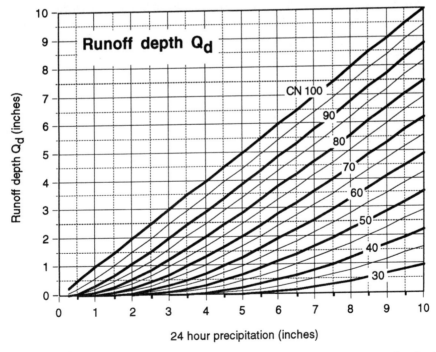

Figure 4.16 SCS storm runoff depth (based on Equation 2-3 of U.S. Soil Conservation Service, 1986, with I_a equal to 0.2S).

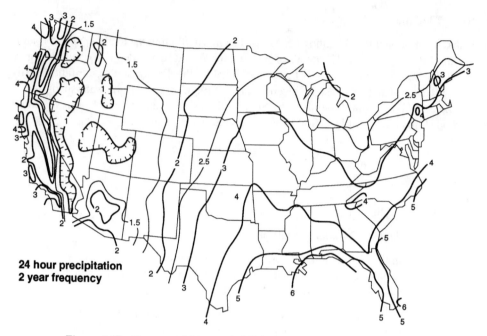

Figure 4.17 Two-year, 24-hour rainfall, in inches (after Hershfield, 1961).

Given the precipitation at a selected recurrence interval and the watershed's curve number CN, depth of runoff during the storm can be found by the preceding equation or from the chart in Figure 4.16. To use the chart, enter from the bottom with the known precipita-

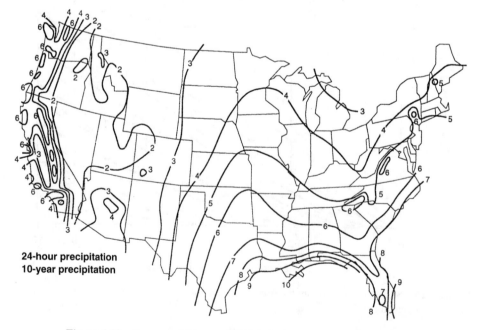

Figure 4.18 Ten-year, 24-hour rainfall, in inches (after Hershfield, 1961).

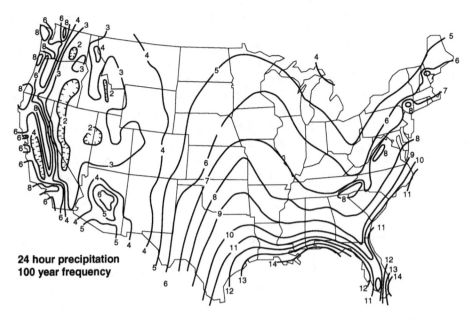

Figure 4.19 Hundred-year, 24-hour rainfall, in inches (after Hershfield, 1961).

tion. Go vertically up to the line for your *CN*. From there move horizontally to the left, to read the depth of runoff Q_d.

For precipitation *P*, the SCS Method always uses 24-hour rainfall, with the intensity of rain within that period assumed to be within a certain distribution. Twenty-four hour rainfall amounts at three recurrence intervals are shown in Figures 4.17 through 4.19. Small-scale maps like those in this book should be used cautiously, particularly in mountainous areas where precipitation can vary greatly in a short distance. Sources of more detailed maps for certain regions are listed in Table 4.2.

In characterizing a drainage area's soil and cover, the SCS Method makes powerful use of the agency's system of classifying soils and mapping them in local soil surveys. Their soil classification system is a hierarchic system analogous to plant taxonomy, in which orders and families are successively divided into more specific categories. The two lowest, most detailed levels of the system are the soil series and the soil type. Within a soil series all soils have identical profiles (sequence of horizons), except that they may have different textures in the A (surface) horizon. There are about 3,000 soil series in the United States; each

TABLE 4.2 Sources of Detailed Data for Precipitation *P* in the SCS Method

Region	Source of Data
Contiguous states west of the 105th meridian	National Oceanic and Atmospheric Administration Atlas 2 (Miller et al., 1973), in separate volumes for the individual states
Alaska	Weather Bureau Technical Paper No. 47 (Miller, 1963)
Hawaii	Weather Bureau Technical Paper No. 43 (U.S Weather Bureau, 1962)
Puerto Rico and Virgin Islands	Weather Bureau Technical Paper No. 42 (U.S. Weather Bureau, 1961)

has an established name, such as Hagerstown or Yolo. The series are described in the soil survey for each locale where they occur. Each series is further divided into soil types—which identify the texture of the A horizon in a specific occurrence of the series, for example, Hagerstown sandy loam or Yolo clay. For characterizing the hydrologic influences of soil, the series is the basic unit of soil classification.

The Hydrologic Soil Group (HSG) is SCS's way of summarizing a soil series' hydrologic effects. SCS has categorized every soil series in the country into four groups, lettered A through D. Group A is the least likely to create runoff; group D is the most likely. Each series is placed in a group according to the descriptions listed in Table 4.3. For a specific site, HSG designations can be obtained by referring to the local SCS soil survey where one is available. If the survey does not specify HSGs, you can look up the soil series in the complete national listing given in Technical Release 55 (U.S. Soil Conservation Service, 1986). If there is no SCS survey at all, you can make an on-site investigation of soil characteristics and compare them with the descriptions in Table 4.3.

Given a soil series' hydrologic soil group, curve number *CN* is derived from a combination of HSG and land use. SCS's curve number summarizes the effects of soil and land use, as does the cover factor in the Rational formula. The name for *CN* comes from the curves in the graph in Figure 4.16, showing the basic SCS relationship between rainfall and runoff. Every combination of hydrologic soil group and land use type has a curve in the graph. Each

TABLE 4.3 Hydrologic Soil Groups Used in the SCS Method (U.S. Soil Conservation Service, 1986, p. A-1)

Group	Description
A	Group A soils have low runoff potential. They have high infiltration rates even when thoroughly wetted. They consist chiefly of deep, well to excessively drained sands or gravels. This group also includes sand, loamy sand and sandy loam that have experienced urbanization but have not been significantly compacted.
B	Group B soils have moderate infiltration rates when thoroughly wetted. They consist chiefly of moderately deep to deep, moderately well- to well-drained soils with moderately fine to moderately coarse textures. This group also includes silt loam and loam that have experienced urbanization but have not been significantly compacted.
C	Group C soils have low infiltration rates when thoroughly wetted. They consist chiefly of soils with a layer that impedes downward movement of water and soils with moderately fine to fine texture. This group also includes sandy clay loam that has experienced urbanization but has not been significantly compacted.
D	Group D soils have high runoff potential. They have very low infiltration rates when thoroughly wetted. They consist chiefly of clay soils with high swelling potential, soils with permanent high water tables, soils with clay pans or clay layers at or near the surface, and shallow soils over nearly impervious material. This group also includes clay loam, silty clay loam, sandy clay, silty clay and clay that have experienced urbanization but have not been significantly compacted.
A/D	The compound classification A/D indicates that the natural soil is in group D because a high water table impedes infiltration and transmission, but following artificial drainage such as with perforated pipe underdrains, the soil's classification is changed to A.

curve has, literally, a curve number. Curve number can vary, theoretically, from 0 to 100 as soil and cover conditions generate increasing amounts of runoff. In practice, curve number is seldom less than 30, and the highest curve number, that for impervious roofs and pavements, is 98. Table 4.4 gives values of *CN* for specific combinations of HSG and land use.

A category of urban surface not covered in SCS's *CN* tables is porous pavement. Until SCS provides a more specific value in a future manual, it can be assumed that areas paved in permeable material are not counted among impervious surfaces and have the same curve number as that of other nearby pervious areas in the watershed.

Composite *CN*s need to be calculated for areas with combinations of different soils or surface covers. To make this calculation, find the *CN*s for the individual types of surface cover, then average them, weighting them by the proportion of the total area they cover.

TABLE 4.4   Values of curve number *CN* in the SCS Method (from Table 2-2 of *Urban Hydrology for Small Watersheds*, *U.S.* Soil Conservation Service, 1986, with some extrapolation). Values in this table assume that antecedent runoff condition (the runoff potential before a storm event) is average, that vegetation is fully established, and that $I_a = 0.2S$

Surface Cover	Hydrologic Soil Group			
	A	B	C	D
Natural areas				
Woods				
No grazing; litter and brush cover the soil	30	55	70	77
Grazed but not burned, some forest litter covers soil	36	60	73	79
Heavy grazing or regular burning destroys litter, brush	45	66	77	83
Desert shrub (saltbush, mesquite, creosote bush, etc.)				
>70% ground cover	49	68	79	84
30 to 70% ground cover	55	72	81	86
<30% ground cover (litter, grass, and brush overstory)	63	77	85	88
Brush, grass, and weeds in arid and semiarid regions				
>70% ground cover	49	62	74	85
30 to 70% ground cover	60	71	81	89
<30% ground cover (litter, grass, and brush overstory)	71	80	87	93
Pinyon and juniper with grass understory				
>70% ground cover	21	41	61	71
30 to 70% ground cover	40	58	73	80
<30% ground cover (litter, grass, and brush overstory)	63	75	85	89
Brush, grass, and weeds in humid regions				
>75% ground cover	30	48	65	73
50 to 75% ground cover	35	56	70	77
<50% ground cover	48	67	77	83
Oak-aspen mountain brush (oak, aspen, bitter brush, etc.)				
>70% ground cover	20	30	41	48
30 to 70% ground cover	38	48	57	63
<30% ground cover (litter, grass, and brush overstory)	56	66	74	79
Sagebrush with grass understory				
>70% ground cover	22	35	47	55
30 to 70% ground cover	37	51	63	70
<30% ground cover (litter, grass, and brush overstory)	52	67	80	85

(continued)

TABLE 4.4 Values of curve number *CN* in the SCS Method (*Continued*)

Surface Cover	Hydrologic Soil Group			
	A	B	C	D
Agricultural areas				
Meadow: grass generally mowed for hay, not grazed	30	58	71	78
Pasture, grassland or range in humid regions				
>75% ground cover and lightly or occasionally grazed	39	61	74	80
50 to 75% grass cover	49	69	79	84
<50% ground cover or heavily grazed with no mulch	68	79	86	89
Woods-grass combination (orchard or tree farm, 50% woods, 50% grass)				
>75% ground cover and lightly or occasionally grazed	32	58	72	79
50 to 75% grass cover	43	65	76	82
<50% ground cover or heavily grazed with no mulch	57	73	82	86
Small grains, contoured and terraced				
High density, ≥20% residue cover	59	70	78	81
Low density, <20% residue cover	61	72	79	82
Small grains, straight rows				
High density, ≥20% residue cover	63	75	83	87
Low density, <20% residue cover	65	76	84	88
Row crops, contoured and terraced				
High density, ≥20% residue cover	62	71	78	81
Low density, <20% residue cover	66	74	80	82
Row crops, straight rows				
High density, ≥20% residue cover	67	78	85	89
Low density, <20% residue cover	72	81	88	91
Farmsteads: buildings, lanes, driveways, and surroundings	59	74	82	86
Fallow				
≥20% residue cover	74	83	88	90
<20% residue cover	76	85	90	93
Bare soil	77	86	91	94
Urban areas				
Lawn, turf				
>75% grass cover	39	61	74	80
50 to 75% grass cover	49	69	79	84
<50% grass cover	68	79	86	89
Desert-shrub landscaping				
Natural desert landscaping, pervious areas only	63	77	85	88
Impervious weed barrier and mulch	96	96	96	96
Newly graded soil, no vegetation or pavements	77	86	91	94
Streets, roads, buildings, structures				
Dirt roads, including right-of-way	72	82	87	89
Gravel roads, including right-of-way	76	85	89	91
Paved road and adjacent swales	83	89	92	93
Impervious roofs and pavements	98	98	98	98

Urban areas, which consist of combinations of pervious and impervious surfaces, involve a special kind of composite *CN*. The composite value is, in principle, a weighted average of the impervious curve number of 98 and the curve number of the surrounding pervious areas. However, the composite value can be reduced where runoff from the impervious surfaces flows across pervious soil and is slowed down and reabsorbed into the soil before reaching the bottom of the watershed.

"Directly connected" impervious surfaces are those from which runoff flows to the drainage system directly via concentrated flow, without reabsorption into the soil. This is most common in densely built-up areas, but it also happens in some other areas where gutters and pipes connect all impervious surfaces into a single drainage system. In a drainage area where impervious surfaces are directly connected, and in any drainage area with more than 30 percent impervious cover, the composite curve number CN_{comp} is found by (equation for Figure 2-3, p. F-1 of *Urban Hydrology for Small watersheds,* U.S. Soil Conservation Service, 1986):

$$CN_{comp} = CN_{perv} + Imp\,(98 - CN_{perv})$$

where

$\quad\quad 98$ = curve number for impervious surfaces

$\quad CN_{perv}$ = curve number for the surrounding pervious surfaces

$\quad\quad Imp$ = portion of the drainage area covered by impervious surfaces, ac/ac

Figure 4.20 shows the results of this equation. To use the chart, enter from the bottom with the curve number of the pervious surfaces. Go vertically up to the line for the percentage of impervious cover, thence to the left to read the composite curve number.

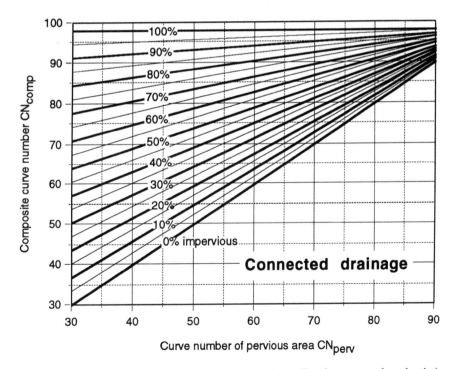

Figure 4.20 Urban composite *CN* with impervious surfaces directly connected to the drainage system, or with total impervious area greater than 30 percent (based on equation for Figure 2-3, given on page F-1 of U.S. Soil Conservation Service, 1986).

"Unconnected" impervious surfaces are those from which runoff spreads as sheet flow over pervious areas before entering the drainage system, so that it can be slowed down and reabsorbed into the soil before continuing to the watershed's outlet. This can occur only in sites with low impervious coverage. In a drainage area where some or all of the impervious surfaces are unconnected, the composite curve number CN_{comp} is given by (equation for Figure 2-4, p. F-1 of *Urban Hydrology for Small* Watersheds, U.S. Soil Conservation Service, 1986):

$$CN_{comp} = CN_{perv} + Imp\,(98 - Cn_{perv})\,(1 - 0.5\,R)$$

where

$$98 = \text{curve number for impervious surfaces}$$

$$CN_{perv} = \text{curve number for the surrounding pervious surfaces}$$

$$Imp = \text{portion of the drainage area covered by impervious surfaces, ac/ac}$$

$$R = \text{ratio of unconnected impervious area to total impervious area, ac/ac}$$

Figure 4.21 shows the results of this equation, with $R = 1$.

If an urban area has a combination of connected and unconnected impervious subareas, substitute the appropriate value of R in the preceding equation, and solve for CN_{comp}.

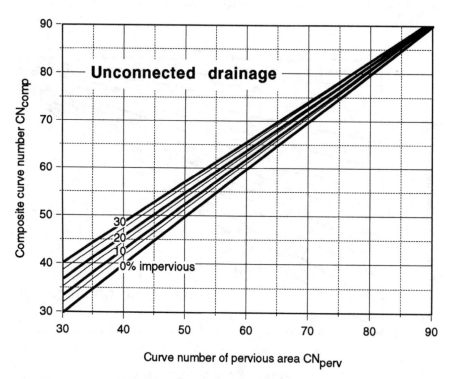

Figure 4.21 Urban composite *CN* with impervious surfaces not connected to the drainage system by concentrated flow (based on equation for Figure 2-4, given on page F-1 of U.S. Soil Conservation Service, 1986, with $R = 1$).

After the depth of runoff Q_d has been found from the basic SCS equation or from Figure 4.16, volume of runoff Q_{vol} is the runoff depth times the watershed area:

$$Q_{vol} = Q_d A_d / 12$$

where

$$Q_{vol} = \text{volume of runoff, af}$$
$$Q_d = \text{depth of runoff, in.}$$
$$A_d = \text{drainage area, acres}$$
$$12 = \text{conversion factor, in./ft}$$

Analytical Exercise

Starting from the equations for urban composite curve number, with $R = 1$, derive algebraically an equation for the ratio between the curve numbers with connected and unconnected impervious surfaces. How different are the two curve numbers? What types of physical hydrologic processes does this difference in curve numbers represent?

Peak Rate of Flow

Finding peak rate of flow q_p is the complicated part of the SCS Method. On a computer, the method involves repeated calculations to add up flows as intensity of rainfall and intensity of runoff vary over time. Manual Methods like those in this book are based on the results of many computer runs by SCS, using its original equations.

SCS (1986) refers to the manual Method presented here as the "graphical peak discharge" method and specifies that it should be applied only to homogeneous individual drainage areas. For watersheds that could be divided into subwatersheds with substantially different curve numbers or times of concentration, you should use either the "tabular hydrograph" method of TR-55 (U.S. Soil Conservation Service, 1986) in manual calculations, or a computer program that routes flows from different subwatersheds together. Such approaches generate hydrographs, telling you how runoff varies from moment to moment during a storm, so that runoff from a number of different subwatersheds can be combined over time.

For the "graphical peak discharge" method, SCS has provided this relatively simple equation (adapted from Equation 4-1 of U.S. Soil Conservation Service TR-55, 1986):

$$q_p = q_u A_d Q_d F_p$$

where

$$q_p = \text{peak rate of flow during a storm event, cfs}$$
$$q_u = \text{unit peak discharge, cfs/ac/in}$$
$$A_d = \text{drainage area, ac}$$
$$Q_d = \text{depth of runoff, in.}$$
$$F_p = \text{factor for ponds and swamps outside the hydraulic length}$$

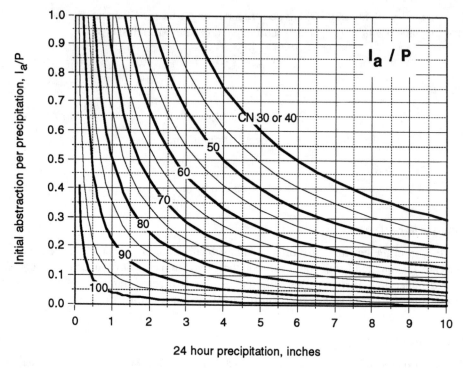

Figure 4.22   Ratio I_a/P as a function of storm precipitation (based on Table 4-1 of U.S. Soil Conservation Service, 1986).

The complicated part of the q_p equation is the factor q_u, unit peak discharge. It takes a few minutes to find its value. After it has been found, the remaining factors in the equation will come quickly.

The equations underlying q_u tend to be complicated and to be ad hoc regressions without direct meaning in physical runoff processes, so we will not bother with them here. Instead we will go, as SCS does in its own publications, directly to charts for the values of the factors. If you want to see the equations, you can find them in Appendix F of TR-55 (U.S. Soil Conservation Service, 1986).

The value of q_u depends on the precipitation's initial abstraction and the watershed's time of concentration.

Find the ratio I_a/P (initial abstraction as a proportion of total precipitation) from the graph in Figure 4.22. Enter the chart from the bottom with the storm precipitation, move up to the line for the drainage area's curve number, thence to the left to read the ratio I_a/P.

For time of concentration, SCS presented a relatively simple estimation method in Technical Publication (TP) 149 (Kent, 1968). This method is considerably simpler than the more recent TR-55 method, so some designers continue to use it in practice. In some jurisdictions the laborious TR-55 method is required; in those instances you can refer to SCS's original publication (U.S. Soil Conservation Service, 1986). The TP-149 method is based on the following equation (adapted from Kent, 1968):

$$t_c = 0.1111 \, L_h^{0.8} \, [(S + 1)^{1.67}] \, / \, G^{0.5}$$

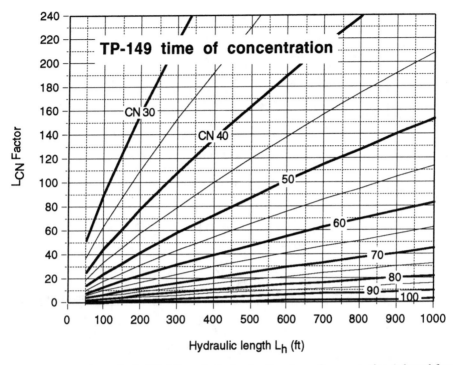

Figure 4.23 L_{CN} factor in the TP-149 method of estimating time of concentration (adapted from Kent, 1968). Time of concentration $tc = L_{CN}/G^{0.5}$, where slope G is in percentage.

where

t_c = time of concentration, minutes

L_h = hydraulic length, ft

S = the retention factor in the basic SCS equation, $S = 1,000/CN - 10$

G = slope along the hydraulic length, percent

To implement this equation in an understandable chart, I rewrote it in the form $t_c = L_{CN}/G^{0.5}$, where the factor L_{CN} combines everything in the equation but $G^{0.5}$. The chart in Figure 4.23 shows the result. To use the chart, enter from the bottom with the hydraulic length L_h. Move up to the line for the watershed's curve number, thence to the left to read the L_{CN} factor. Then apply the factor to find time of concentration, $t_c = L_{CN}/G^{0.5}$.

If the hydraulic length passes over surfaces with substantially different slopes or CN values, find the travel time over each homogeneous portion of the hydraulic length separately. The total time of concentration is the sum of the times across the various surfaces.

The application of I_a/P and time of concentration to find q_u depends on the short-term intensity of rainfall. SCS assumes that the intensity within a 24-hour storm is distributed from hour to hour differently in different climatic regions. Figure 4.24 shows the distributions in four regional storm types. SCS's assumed distribution contains the correct intensity for any duration less than 24 hours; for example, the most intense hour during the 10-year,

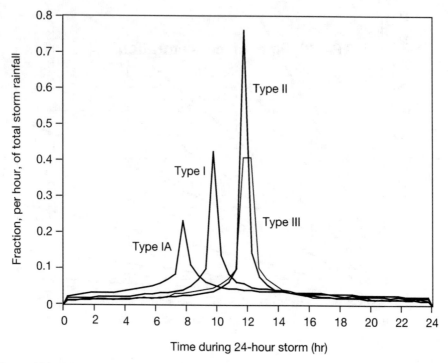

Figure 4.24 SCS's assumed distribution of rainfall during a 24-hour storm in four regional storm types (based on Debo and Reese, 1995, p. 232).

24-hour storm has the intensity of the 10-year, 1-hour storm. The map in Figure 4.25 shows the regions where the four storm distribution types occur. In most of the country, with Type II, design storms are apparently brief but intense summer thunderstorms. In region III they are coastal storms and hurricanes, and on the West Coast they are long, slow, misty drizzles off the Pacific Ocean.

Find the graph of unit peak discharge q_u for your rainfall distribution among Figures 4.26 through 4.29. Enter the appropriate graph from the bottom with the time of concentration. Go up to the line for your I_a/P ratio, thence horizontally to the left to read unit peak discharge q_u, the peak flow per acre of watershed per inch of runoff. Use the value in SCS's equation for q_p.

The pond-and-swamp factor F_p is intended to take into account the slowing of runoff as it passes through topographic basins and slow-draining areas. It applies to watersheds where pond and swamp areas are spread throughout a watershed and do not all lie along the hydraulic length, and so were not fully considered in the estimate of time of concentration. In watersheds that do not have such ponds or swamps, F_p is equal to 1 and can be disregarded. For watersheds that have such ponds or swamps, find F_p from the graph in Figure 4.30 and use it in the equation for q_p.

SCS Method Exercises

Exercises 4.3 and 4.4 find SCS runoff volume and peak rate before and after development, so they can be compared. The results can also be compared with those derived from the

Figure 4.25   SCS rainfall distributions (after Figure B-2 of U.S. Soil Conservation Service, 1986).

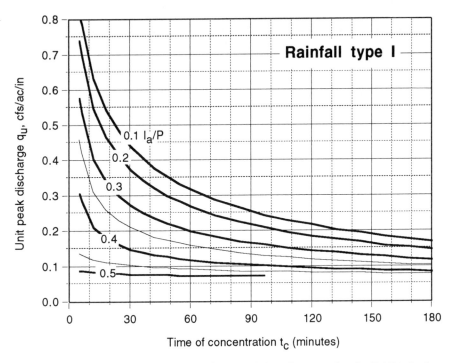

Figure 4.26   Unit peak discharge q_u in rainfall Type I (based on equation for Exhibit 4, given on page F-1 of U.S. Soil Conservation Service, 1986).

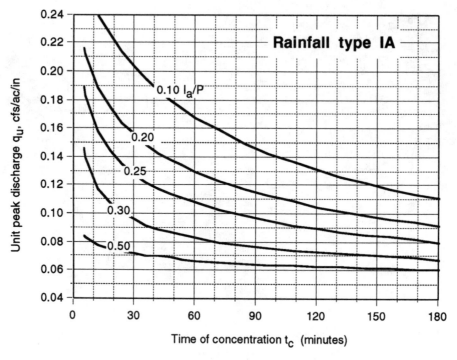

Figure 4.27 Unit peak discharge q_u in rainfall Type IA (based on equation for Exhibit 4, given on page F-1 of U.S. Soil Conservation Service, 1986).

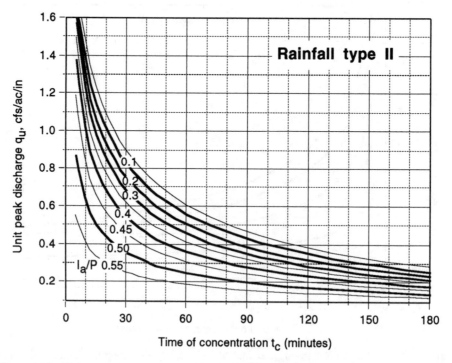

Figure 4.28 Unit peak discharge q_u in rainfall Type II (based on equation for Exhibit 4, given on page F-1 of U.S. Soil Conservation Service, 1986, and some extrapolation).

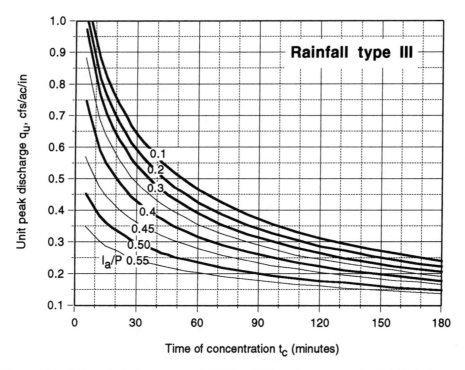

Figure 4.29 Unit peak discharge q_u in rainfall Type III (based on equation for Exhibit 4, given on page F-1 of U.S. Soil Conservation Service, 1986, and some extrapolation).

Rational Method, found in Exercises 4.1 and 4.2. These exercises assume that the drainage area before development is homogeneous, and that after development it is a mosaic of pervious turf and impervious roofs and pavements. The results are used for design in Chapters 6 through 10.

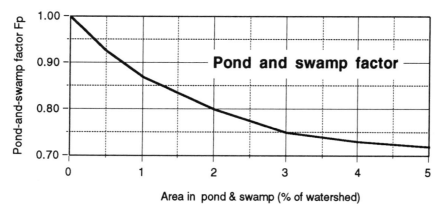

Figure 4.30 SCS's pond-and-swamp factor F_p (based on Table 4-2 of U.S. Soil Conservation Service, 1986).

Exercise 4.3   SCS Method, Before Development

	Site 1		Site 2	

Runoff volume

24 hour rainfall P (from Figures 4.17-4.19)	= _____	in.	_____	in.
Hydrologic soil group HSG (from soil survey)	= _____		_____	
Curve number (from Table 4.4)	= _____		_____	
Runoff depth Q_d (from Figure 4.16)	= _____	in.	_____	in.
Drainage area A_d (from site map)	= _____	ac	_____	ac
Runoff volume Q_{vol} before $= q_d A_d / 12$	= _____	af	_____	af

Peak rate of flow

Ratio I_d/P (from Figure 4.22)	= _____		_____	
Hydraulic length L_h (from site map)	= _____	ft	_____	ft
L_{CN} factor (from Figure 4.23)	= _____		_____	
Slope along hydraulic length G (from site map)	= _____	%	_____	%
Time of concentration t_c $= L_{CN} / G^{0.5}$	= _____	min	_____	min
Rainfall distribution type (from Figure 4.25)	= _____		_____	
Unit peak discharge q_u (from Figures 4.26–4.29)	= _____	cfs/ac/in.	_____	cfs/ac/in.
Pond-and-swamp factor F_p (from Figure 4.30)	= _____		_____	
Peak rate of flow q_p before $= q A_d Q_d F_p$	= _____	cfs	_____	cfs

Summary of Process

Runoff volume

1. Obtain 24-hour rainfall in inches, at a selected recurrence interval, from Figures 4.17 through 4.19, one of the sources listed in Table 4.2, or locally available data.

2. Obtain hydrologic soil group HSG from the local SCS soil survey that applies to your site, or by investigating soils on-site and applying SCS's descriptions found in Table 4.3.

Exercise 4.4   SCS Method, After Development

	Site 1	Site 2

Runoff volume

24 hour rainfall P
(from Figures 4.17-4.19)   = _____ in.   _____ in.
Hydrologic soil group HSG
(from soil survey)   = _____   _____
Impervious area
(from site map)   = _____ %   _____ %
Curve number of pervious surface
(from Table 4.4)   = _____   _____
Composite curve number
(from Fig. 4.20 or 4.21)   = _____   _____
Runoff depth Q_d
(from Figure 4.16)   = _____ in.   _____ in.
Drainage area A_d
(from site map)   = _____ ac   _____ ac
Runoff volume Q_{vol} after
$= q_d A_d / 12$ = _____ af   _____ af

Peak rate of flow

Ratio I_a/P
(from Figure 4.22)   = _____   _____
Hydraulic length on turf
(from site map)   = _____ ft   _____ ft
L_{CN} factor for turf
(from Figure 4.23)   = _____   _____
Slope on turf G
(from site map)   = _____ %   _____ %
Travel time on turf t_1
$= L_{CN} / G^{0.5}$ = _____ min   _____ min
Hydraulic length on pavement
(from site map)   = _____ ft   _____ ft
L_{CN} factor for pavement
(from Figure 4.23)   = _____   _____
Slope on pavement G
(from site map)   = _____ %   _____ %
Travel time on pavement t_2
$= L_{CN} / G^{0.5}$ = _____ min   _____ min
Time of concentration t_c
$= t_1 + t_2$ = _____ min   _____ min
Rainfall distribution type
(from Figure 4.25)   = _____   _____
Unit peak discharge q_u
(from Figures 4.26–4.29)   = _____ cfs/ac/in.   _____ cfs/ac/in.
Pond-and-swamp factor F_p
(from Figure 4.30)   = _____   _____
Peak rate of flow q_p after
$= q_p A_d Q_d F_p$ = _____ cfs   _____ cfs

3. Obtain curve number CN from Table 4.4. For a drainage area with more than one HSG or land use, compute the weighted average CN for the area. For an urban drainage area containing impervious surfaces, obtain the percentage of impervious cover by scaling a site map. If the impervious cover is less than 30 percent, determine from site plans or on-site observation whether the impervious surfaces are directly connected to the drainage system. Obtain the CN of the pervious cover from Table 4.4. Find the composite urban curve number from Figure 4.20 or 4.21.

4. Find runoff depth Q_d in inches from the basic SCS equation or the graph in Figure 4.16.

5. Estimate drainage area A_d in acres from a site map.

6. Compute volume of flow Q_{vol} in af by the equation, $Q_{vol} = Q_d A_d / 12$.

Peak Rate of Flow

7. Obtain ratio of initial abstraction to precipitation I_a/P from the graph in Figure 4.22.

8. Obtain hydraulic length L_h and its slope G from a site map.

9. Obtain time of concentration of the drainage area. First find the L_{CN} factor from Figure 4.23, then calculate time of concentration $t_c = L_{CN} / G^{0.5}$. If the hydraulic length passes through areas with substantially different slopes or CN values, find the time of travel for each homogeneous portion; the total time of concentration is the sum of the travel times through the various portions.

10. Find storm rainfall distribution type from the map in Figure 4.25.

11. Obtain unit peak discharge q_u (cfs/ac/in.) from the appropriate graph in Figures 4.26 through 4.29.

12. Obtain pond-and-swamp factor F_p (no units) from the graph in Figure 4.30.

13. Compute peak rate of runoff q_p in cfs by the equation, $q_p = q_u A_d Q_d F_p$.

Discussion of Results

1. Which method, Rational or SCS, produces the larger peak rate of runoff q_p on each site, before and after development? What specific factors in the methods led to these results?

2. Considering the Rational and SCS runoff estimation methods, would you select one method over another? On what grounds?

3. Outline the volume of one cubic foot with your hands. Do you think either of the runoff estimation methods is sufficiently precise to give you answers in fractions of a cubic foot? To what digit do you think you should round off your calculations of Q_{vol} and q_p?

4. Which site has the larger volume Q_{vol} and peak rate q_p of runoff before development? Why?

5. On which site does the volume of runoff Q_{vol} increase by the greatest amount (af) when the land is developed? Why? On which does it increase by the greatest proportion (Q_{vol} after ÷ Q_{vol} before)? Why?

6. With the SCS Method, on which site does the peak rate of runoff q_p increase by the greatest amount (cfs) when the land is developed? Why? On which does it increase by the greatest proportion (q_p after ÷ q_p before)? Why?

7. In terms of Q_{vol} and q_p, does development cause the two sites to become hydrologically more similar, or more different? Why?

8. If you were planning a region's future land use for the purpose of preventing environmental damage to streams, on what types of soils, slopes, and land cover would you propose development with a high proportion of impervious cover? Why?

9. How could you alter a development's land use type, layout, or construction materials to reduce the volume of runoff that it generates? Identify several specific ways.

10. Given that a development will generate a certain total volume of runoff, how could you alter the layout so as to reduce the volume that reaches a given place in the drainage system? Be specific. What effects would such alterations have on the runoff reaching other parts of the drainage system?

11. Find the proportion of rainfall that becomes runoff by dividing the SCS runoff depth Q_d by the 24-hour storm rainfall, before and after development. What is the difference in the runoff/rainfall ratio between existing and proposed conditions? What types of physical processes in the watershed bring about this difference? If some of the rain did not discharge as storm runoff, where did it go? Will it ever reach the streams?

HOW A COMPUTER CAN HELP

Commercially available software assists practitioners in estimating storm runoff using established rainfall-runoff models. Systematic analyses of large, complex watersheds that are not feasible by hand can be done on a computer. By using a model to go rapidly through a number of calculations, a practitioner can find out the effects of different estimation assumptions and design alternatives.

One type of computer analysis is flood routing. As a given volume of water moves downstream in a flood wave through a network of channels and reservoirs, it is attenuated by temporary storage at each step. On a hydrograph, the time base is lengthened and the peak flow is reduced, until the next tributary adds more runoff. Given the flows generated by upstream drainage areas, the analysis combines converging flows and routes the calculations downstream.

Different commercially available programs use different hydrologic estimation methods, require different kinds of input information, and produce different kinds and amounts of results. Most software packages contain several modules for distinct steps; such as inputting watershed data, estimating runoff, routing downstream flows, and evaluating different management alternatives. With time, some programs are becoming integrated with CAD and GIS, and useful for rapid investigation of different design alternatives.

Although many of the hydrologic models underlying computer software originated with agencies such as SCS and the U.S. Environmental Protection Agency (EPA), commercial vendors have integrated procedures from different sources, placed interfaces on them, and facilitated input and output. The vendors are all trying to make it in a competitive market; their products and services are changing rapidly. To find out what is currently available, look for the latest products at professional conferences and in professional magazines.

In using hydrologic software, exercise care and common sense to make sure the computer is doing what you think it is, and that it is coming up with a reasonable answer. Review data that you have been inputting. Check data that do not look familiar or that do not make sense. For instance, are the relative sizes of drainage areas consistent with the way they appear on a map? Is the percentage of impervious cover consistent with the general intensity of land use? Be sure to review the model's results. Does development increase runoff rate and volume as it is ordinarily expected to? Does time of concentration increase or decrease with development, and can you tell why? Do flows and times increase downstream in a logical pattern?

In practice, your knowledge and judgment should never be replaced by a programmed model or its results. There is no use trying to use a model unless you know stormwater hydrology first. A sophisticated tool does not protect an ignorant or careless user. The responsibility for being in control of your project and coming to the right conclusion is always yours. Your personal ability to carry out underlying equations by hand will always affirm your ability to command any hydrologic software, of any degree of sophistication or complexity.

SUMMARY AND COMMENTARY

You have not yet designed anything. When you have estimated runoff, you have only modeled a natural process. You have made an estimate of what will happen if a certain rain falls on your site.

The relative accuracy of different runoff estimation methods has been the subject of long arguments. Many of those arguments have in fact been unresolvable, because they have taken place without the benefit of actual measured on-site runoff data. Some people have argued that the SCS Method is inherently more accurate than the Rational because it is more "sophisticated," in the sense that it is more complex and takes a greater number of factors explicitly into account. But sophistication does not equal accuracy. It has also been argued that the SCS Method embodies a superior theory of the mechanisms by which runoff is generated on the ground. But a superior theory (if there really is one) does not in itself create superior accuracy, any more than superior sophistication does.

Accuracy is a positive relationship between the results of a method—the estimated volume and rate of runoff—and what actually happens in real drainage areas, during real rainstorms. A high level of accuracy means that the results of a model closely approximate real flows.

Accurate results can be assured only when the runoff model you are using has been calibrated to actual local conditions. Calibrating a model requires acquisition of local data. Precipitation and stream gauging stations have to be set up for a period of at least three to five years in a watershed with stable land use. Streamflow and rainfall data from storm events would then be collected. Using these data, a selected estimation method can be calibrated or manipulated to yield estimates of runoff similar to that which has been observed in the field. The calibrated model can then be used in other nearby areas that have hydrologic characteristics similar to those of the watersheds where gauging took place for estimating runoff with known accuracy.

The SCS Method is more amenable to such calibration than the Rational formula, because it has more variables describing drainage area characteristics and runoff behavior. Each of these variables can be changed or calibrated to describe local conditions relatively precisely.

Too frequently—in fact, routinely—we have to design facilities that are going to be built right away, in locales that have not been monitored even for three years. In these cases methods must be used to make estimates of natural processes based on general knowledge, with the understanding that site-specific accuracy is not a definable issue.

Which model to choose for a specific project is often a matter of many considerations other than accuracy. The SCS Method tends to be more consistent in its application. Depending on local conditions, one method or the other may be more conservative, in the sense that its use usually ends up requiring facilities with greater capacities. Local agencies that review development plans often accept only one particular method for a given type of application.

Snowmelt is a major component of runoff in some regions, such as the mountainous parts of the Pacific Northwest. In those places runoff models can be used that take into account snowmelt as well as rainfall. The methods discussed in this chapter do not take snowmelt into account. If you work in such a region you can find out about locally preferred models by referring to hydrologic guides published by local agencies.

REFERENCES

Aron, Gert, David J. Wall, Elizabeth L. White, and Christopher N. Dunn, 1987, Regional Rainfall Intensity-Duration-Frequency Curves for Pennsylvania, *Water Resources Bulletin* vol. 23, no. 3, pp. 479ñ485.

Debo, Thomas N., and Andrew J. Reese, 1995, *Municipal Storm Water Management*, Boca Raton: Lewis.

Debo, Thomas N., and George E. Small, 1989, Hydrologic Calibration: The Forgotten Aspect of Drainage Design, *Public Works*, February, pp. 58–59.

Frederick, Ralph H., V. A. Myers, and E. P. Auciello, 1977, *Five to 60 Minute Precipitation and Frequency for the Eastern and Central United States*, Technical Memorandum NWS HYDRO-35, Silver Spring, Md.: National Weather Service, Office of Hydrology.

Hershfield, David M., 1961, *Rainfall Frequency Atlas of the United States*, Technical Paper 40, Washington: U.S. Department of Commerce, Weather Bureau.

Kent, K. M., 1968, *A Method for Estimating Volume and Rate of Runoff in Small Watersheds*, Technical Publication 149, Washington: U.S. Soil Conservation Service.

Kuichling, Emil, 1889, The Relation Between the Rainfall and the Discharge of Sewers in Populous Districts, *Transactions of the American Society of Civil Engineers* vol. 20, January, p. 1–60.

Leopold, Luna B., 1974, *Water, A Primer*, San Francisco: Freeman.

Miller, John F., 1963, *Probable Maximum Precipitation and Rainfall-Frequency Data for Alaska*, Technical Paper No. 47, Washington: U.S. Weather Bureau.

Miller, John F., R. H. Frederick, and R. J. Tracey, 1973, *Precipitation-Frequency Atlas of the Western United States*, Atlas 2, Washington: National Oceanic and Atmospheric Administration, (in separate volumes for individual states).

U.S. Federal Aviation Agency, 1965, *Airport Drainage*, AC 150/5320-5A, Washington: U.S. Federal Aviation Agency.

U.S. Federal Highway Administration, 1973, *Design of Roadside Drainage Channels*, Hydraulic Design Series No. 4, Washington: U.S. Federal Highway Administration.

U.S. Soil Conservation Service, 1972, *National Engineering Handbook*, Section 4, Hydrology, SCS/ENG/NEH-4, Washington: U.S. Soil Conservation Service.

U.S. Soil Conservation Service, 1986, *Urban Hydrology for Small Watersheds*, Technical Release 55, second edition, Washington: U.S. Soil Conservation Service.

U.S. Weather Bureau, 1955, *Rainfall Intensity-Duration-Frequency Curves for Selected Stations in the United States, Alaska, Hawaiian Islands, and Puerto Rico*, Technical Paper No. 25, Washington: U.S. Weather Bureau.

U.S. Weather Bureau, 1961, *Generalized Estimates of Probable Maximum Precipitation and Rainfall-Frequency Data for Puerto Rico and Virgin Islands*, Technical Paper No. 42, Washington: U.S. Weather Bureau.

U.S. Weather Bureau, 1962, *Rainfall-Frequency Atlas of the Hawaiian Islands*, Technical Paper No. 43, Washington: U.S. Weather Bureau.

CHAPTER 5

WATER BALANCE

A water balance, like a storm runoff estimation, establishes volumes and rates of flow. It is a modeling of a natural process.

But unlike an estimate of storm runoff, a water balance is a complete inventory of the hydrology of a landscape. It is complete in the sense that it is extended over time, aggregating the effects of many small and large storms and showing the changes with the seasons. It is complete also in the sense that it characterizes the entire regime of where water is and what it is doing; it presents a unified view of the moisture environment. For these reasons it has long been a prominent concept in geography, where it is used to predict summary characteristics of biotic communities, stream flow, and human adaptation (Mather, 1978).

Storm runoff and water balance estimations supplement each other as tools for evaluation and design. Storm runoff estimation is needed to protect against, control, and utilize runoff during individual storm events; water balance estimation shows the effects of land use and stormwater control on the local ecosystem. The water balance has specific design applications in extended detention, infiltration, and water harvesting.

CONCEPT AND FORMAT

The underlying principle of the water balance is the change-of-volume concept that was introduced in Chapter 3. The following equation applies to a landscape, or to any component of a landscape, during any given increment of time:

$$\Delta storage = inflow - outflow$$

The term *balance* in the expression *water balance* refers to the conceptual necessity for this equation to balance. Any difference between inflow and outflow must be accounted for

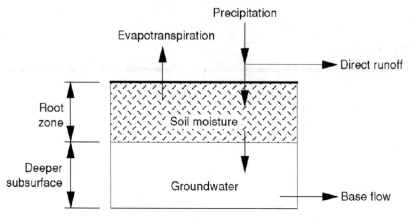

Figure 5.1 Thornthwaite's water balance concept.

by a change in the amount of water stored somewhere in the landscape, whether in snow-pack, soil moisture, groundwater, or surface water bodies.

The inflow to a landscape always includes precipitation. On some sites there are additional inflows, such as stream flow and artificial irrigation. This chapter will work with only the precipitation inflow; for sites where additional inflows are present, appropriate terms can be added to precipitation.

The outflow from a landscape always includes evapotranspiration, direct runoff, and base flow. On some sites there are additional outflows, such as withdrawals for water supply; where necessary, appropriate terms to express additional outflows can be added to the analysis.

The water balance concept was developed by the American climatologist C. W. Thornthwaite in the mid-twentieth century (Thornthwaite, 1948; Thornthwaite and Mather, 1955, 1957). The concept is illustrated in Figure 5.1, where soil moisture and groundwater are places where water is stored, and the arrows represent flows through them. In the sequence of flows, the results of each step set the stage for the next step. Some of the arriving precipitation is diverted into direct runoff (in cold regions, precipitation is first stored temporarily in snowpack). The water that enters the root zone of the soil is subject to evapotranspiration, after which the rest percolates into the deeper subsurface. Groundwater discharges slowly as base flow.

The Thornthwaite water balance applies average monthly data to this sequence of flows and storages. Computer models developed since Thornthwaite's time have produced sophisticated "continuous simulation" water balances that apply large quantities of data at daily or shorter time increments to numerous soil layers. But Thornthwaite's modest concept retains the advantages of simplicity and conciseness in arriving at valid results. As a tool for learning, it provides a basis for understanding more complex continuous simulation models.

Thornthwaite's water balance format arranges monthly data in a "bookkeeping" or "spreadsheet" format. Complete examples from Boston and Los Angeles are presented in Tables 5.1 and 5.2. The results from these two sites will serve as examples when you do calculations for your own site.

Some of the results for Boston are graphed in Figure 5.2. Boston has precipitation year-round. But evapotranspiration (*Et*) varies from zero in the winter to a peak in the summer. The result is high base flow in the winter when there is excess precipitation, even though

TABLE 5.1 Water balance for a site in Boston, Massachusetts. All values are in inches except temperature, which is in °F. The soil's available water-holding capacity is 8 inches. SCS curve number is 65. Heat index I is 45.55. Latitude is 42.3°.

	Jan.	Feb.	Mar.	Apr.	May	June	July	Aug.	Sept.	Oct.	Nov.	Dec.
Climatic data												
Precipitation P	3.58	3.40	3.84	3.57	3.25	3.16	3.15	3.57	3.25	3.22	3.89	3.67
Temperature	28.9	29.1	36.9	46.9	57.7	67.0	72.6	70.7	64.0	54.2	43.5	32.6
Snow storage												
Beginning snow depth	0.00	3.58	6.98	3.56	0.00	0.00	0.00	0.00	0.00	0.00	0.00	0.00
P to snowpack	3.58	3.40	0.00	0.00	0.00	0.00	0.00	0.00	0.00	0.00	0.00	0.00
Snowmelt	0.00	0.00	3.42	3.56	0.00	0.00	0.00	0.00	0.00	0.00	0.00	0.00
Change in snow depth	3.58	3.40	−3.42	−3.56	0.00	0.00	0.00	0.00	0.00	0.00	0.00	0.00
Net precipitation	0.00	0.00	7.26	7.13	3.25	3.16	3.15	3.57	3.25	3.22	3.89	3.67
Direct runoff												
Direct Runoff	0.00	0.00	0.40	0.39	0.13	0.12	0.12	0.15	0.13	0.13	0.17	0.16
Potential evapotranspiration Et_o												
Et_o at equator	0.00	0.00	0.34	1.30	2.52	3.66	4.39	4.14	3.29	2.11	0.95	0.03
Et_o at site / Et_o at equator	0.78	0.88	0.99	1.11	1.22	1.28	1.25	1.16	1.04	0.92	0.81	0.76
Et_o at site	0.00	0.00	0.33	1.44	3.07	4.67	5.50	4.82	3.43	1.95	0.77	0.02
Soil moisture and Et												
Potential moisture increment	3.58	3.40	3.10	1.73	0.05	−1.63	−2.47	−1.40	−0.31	1.15	2.95	3.49
Cumulative potential gain	11.17	14.57	17.67	19.40	19.46	0.00	0.00	0.00	0.00	1.15	4.09	7.59
Cumulative potential loss	0.00	0.00	0.00	0.00	0.00	−1.63	−4.10	−5.50	−5.81	0.00	0.00	0.00
Soil moisture storage SM	8.00	8.00	8.00	8.00	8.00	6.52	4.79	4.02	3.87	5.02	7.96	8.00
ΔSM storage	0.00	0.00	0.00	0.00	0.00	−1.48	−1.73	−0.77	−0.15	1.15	2.95	0.04
Evapotranspiration	0.00	0.00	0.33	1.44	3.07	4.52	4.76	4.19	3.28	1.95	0.77	0.02
Groundwater and base flow												
Groundwater increment	3.58	3.40	3.10	1.73	0.05	0.00	0.00	0.00	0.00	0.00	0.00	3.46
Groundwater storage	5.32	6.06	6.13	4.80	2.45	1.23	0.61	0.31	0.15	0.08	0.04	3.48
Base flow	2.66	3.03	3.07	2.40	1.23	0.61	0.31	0.15	0.08	0.04	0.02	1.74
Detention	2.66	3.03	3.07	2.40	1.23	0.61	0.31	0.15	0.08	0.04	0.02	1.74

TABLE 5.2 Water balance for a site in Los Angeles, California. All values are in inches except temperature, which is in °F. The soil's available water-holding capacity is 8 inches. SCS curve number is 65. Heat index I is 79.85. Latitude is 34.0°.

	Jan.	Feb.	Mar.	Apr.	May	June	July	Aug.	Sept.	Oct.	Nov.	Dec.
Climatic data												
Precipitation P	3.04	3.03	2.51	1.08	0.30	0.06	0.01	0.03	0.20	0.51	1.38	2.63
Temperature	55.9	56.9	58.5	60.8	63.3	67.0	71.5	72.3	70.7	66.5	62.1	57.6
Direct runoff												
Direct Runoff	0.11	0.11	0.08	0.00	0.00	0.00	0.00	0.00	0.00	0.00	0.00	0.09
Potential evapotranspiration Et_o												
Et_o at equator	1.56	1.67	1.87	2.17	2.51	3.06	3.80	3.93	3.66	2.99	2.34	1.76
Et_o at site / Et_o at equator	0.84	0.91	1.00	1.08	1.16	1.20	1.19	1.13	1.03	0.95	0.87	0.82
Et_o at site	1.31	1.53	1.86	2.34	2.92	3.68	4.51	4.43	3.79	2.83	2.03	1.44
Soil moisture & Et												
Potential moisture increment	1.62	1.39	0.57	−1.26	−2.62	−3.62	−4.50	−4.40	−3.59	−2.32	−0.65	1.10
Cumulative potential gain	2.72	4.10	4.68	0.00	0.00	0.00	0.00	0.00	0.00	0.00	0.00	1.10
Cumulative potential loss	0.00	0.00	−3.83	−5.09	−7.72	−11.3	−15.8	−20.2	−23.8	−26.1	−26.79	0.00
Soil moisture storage SM	3.00	4.39	4.96	4.23	3.05	1.94	1.10	0.64	0.41	0.30	0.28	1.38
ΔSM storage	1.62	1.39	0.57	−0.72	−1.18	−1.11	−0.83	−0.47	−0.23	−0.10	−0.02	1.10
Evapotranspiration	1.31	1.53	1.86	1.80	1.48	1.17	0.84	0.50	0.43	0.61	1.40	1.44
Groundwater and base flow												
Groundwater increment	0.00	0.00	0.00	0.00	0.00	0.00	0.00	0.00	0.00	0.00	0.00	0.00
Groundwater storage	0.00	0.00	0.00	0.00	0.00	0.00	0.00	0.00	0.00	0.00	0.00	0.00
Base flow	0.00	0.00	0.00	0.00	0.00	0.00	0.00	0.00	0.00	0.00	0.00	0.00
Detention	0.00	0.00	0.00	0.00	0.00	0.00	0.00	0.00	0.00	0.00	0.00	0.00

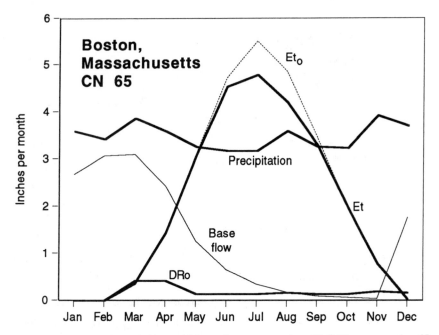

Figure 5.2 Water balance in Boston, Massachusetts, on a site with SCS curve number 65.

some winter precipitation is held in snowpack. Direct runoff becomes high in the spring, when snowmelt multiplies the amount of water available for runoff. The mesic soil supports hardwood forests; streams and wetlands are abundant.

Figure 5.3 shows some of the results for Los Angeles. Los Angeles has a dry climate; in the summer its precipitation is almost nonexistent. Summer evapotranspiration is prevented from reaching zero only because moisture stored in the soil in previous months is drawn down. Both direct runoff and base flow are very low, even in the wet season, because there is so little excess moisture to support them. The drought-stressed soil supports only coastal scrub and fire-prone chaparral.

The climatic data required for the Thornthwaite water balance are monthly average temperature and precipitation, which are monitored at weather stations all over the world. Data for stations in the United States can be obtained from the National Climatic Data Center of the National Oceanic and Atmospheric Administration (NOAA). The Center's address is Federal Building, Asheville, North Carolina 28801. The Center publishes summary statistics for all stations in each state in a monthly and annual series called *Climatological Data*. Data for specific stations are published in a monthly and annual series entitled *Local Climatological Data*. Commercial outfits periodically republish the same data, as does, for example, the Gale Research Company in its *Weather Almanac* and *Weather of U.S. Cities*.

The following sections take you through a Thornthwaite water balance step by step, beginning with the fall of precipitation and its potential storage in a snowpack. The mathematics behind the steps are covered in Thornthwaite and Mather (1955, 1957), Dunne and Leopold (1978, p. 238–248), and Ferguson (1994, p. 99–109, and 1996). If you choose to look up the equations and apply them in a spreadsheet, set the spreadsheet for iterative calculations, because the results of each month are the input for each following month.

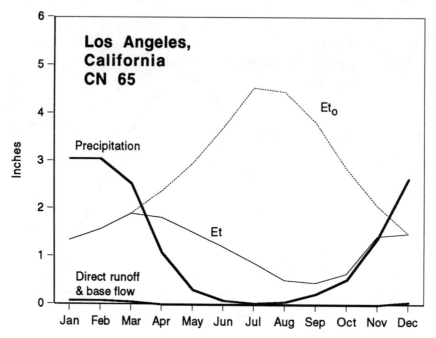

Figure 5.3 Water balance in Los Angeles, California, on a site with SCS curve number 65.

Analytical Exercises

1. Note the recurrence interval and duration of the design storm that you used in Exercise 4.3. Apply appropriate conversion factors to express both in days. Divide the duration by the recurrence interval to find the proportion of total elapsed time occupied by the storm; express the result as a percentage. From the perspective that this proportion gives you, to what relative degrees and in what ways are the design storm and the total long-term rainfall important to the human and natural life of the site?

2. Find the average annual precipitation at your site. Calculate how much rain falls between occurrences of the design storm by multiplying recurrence interval (years) by the annual rainfall (inches per year). If the result is a large number, express it in feet. Get a tangible feel for this quantity by comparing it with your own height, and with the height of the buildings where you live and work. Find the proportion of the site's total rainfall contributed by the design storm by dividing the storm rainfall (inches) by the total rainfall during the recurrence interval (inches). From the perspective that this information gives you, to what relative degrees and in what ways are the design storm and the total long-term rainfall important to the human and natural life of the site?

3. How would your answers to the preceding questions be different if the total rainfall at your site were equal to the U.S. average of 30 inches per year? Where in the United States does rainfall of 30 inches per year occur? How do the seasonal climate, vegetation, and economic base of that region differ from those in your locale?

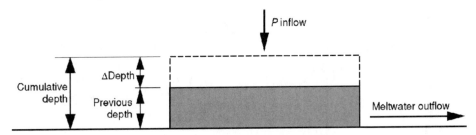

Figure 5.4 Snow storage routing.

SNOW STORAGE

In cold regions, water may be stored as snow for a while before it can participate in the rest of the water balance. The snowpack is a reservoir that captures precipitation, concentrating its release in the time of snowmelt. In warm regions where snow does not significantly accumulate, this preliminary step of estimating snow storage can be skipped, because it makes no difference to the rest of the water balance.

Snowpack inflow and outflow are illustrated in Figure 5.4. The depth of snow is expressed as the equivalent depth of liquid water, the same as precipitation.

When the temperature is adequately low, incoming precipitation adds to the snowpack; it is reflected in an increase in the depth of the snowpack and a decrease in the precipitation that reaches the soil. It is assumed that for each degree Fahrenheit the monthly average temperature falls below 32°, 10 percent of the monthly precipitation is stored as snow.

When the temperature is warm enough, melt water discharges from the snowpack it is reflected in a decrease in the depth of the snowpack, and an increase in the amount of liquid water that reaches the soil. It is assumed that for each degree the temperature rises above $32°$, 10 percent of the accumulated snow melts.

The difference between addition and melt produces each month's change in snow depth (ΔD_{snow}). When ΔD_{snow} is subtracted from each month's "new" precipitation, the difference is the net precipitation available to the soil; it is used in place of "new" precipitation in the rest of the water balance. The cumulative snow depth is carried from each cold month to the next, until the entire snowpack melts.

Snow Storage Exercise

Exercise 5.1 finds snow storage and its outflow. Snow storage is the same before and after development. Fill in the blanks with your data and calculations for the same sites used in Exercises 4.1 through 4.4. The net precipitation P_{net} found in this exercise is used in place of precipitation P in all subsequent water balance calculations. The results of all water balance estimates are used for design in Chapters 8, 9, and 10.

Summary of Process

1. Obtain monthly average precipitation P in inches/month, and monthly average temperature T in °F. Start your calculations at the beginning of a month late in the warm season when the snowpack depth D_{snow} equals zero. Then do the following steps for

Exercise 5.1 Snow storage. All values are in inches except temperature, which is in °F. Complete this exercise for both Site 1 and Site 2. Snow storage is the same before and after development.

		Jan.	Feb.	Mar.	Apr.	May	June	July	Aug.	Sept.	Oct.	Nov.	Dec.
					Climatic data								
Precipitation P (from data source)	=												
Temperature T (from data source)	=												
					Snow storage								
Beginning snow depth D_{snow} = previous D_{snow} + previous ΔD_{snow}	=												
P to snow (if $T < 32$) = $0.1\,P\,(32 - T)$	=												
Snow melt (if $T > 32$) = $0.1\,D_{snow}\,(T - 32)$	=												
Change in snow depth ΔD_{snow} = P to snow − snowmelt	=												
Net precipitation P_{net} = $P - \Delta snow$	=												

each subsequent month. The results of each month's calculations determine the snow-pack depth at the beginning of the next month.

2. If $T < 32°$, find precipitation contributing to snowpack from

$$P \text{ to snow} = 0.1 \, P \, (32 - T)$$

3. If $T > 32°$, find snowpack melt from

$$\text{Snow melt} = 0.1 \, D_{snow} \, (T - 32)$$

4. Find the change in snowpack during the month from

$$\Delta D_{snow} = P \text{ to snow} - \text{snowmelt}$$

5. Find the net precipitation P_{net} available to the soil from

$$P_{net} = P - \Delta D_{snow}$$

6. Move to the next month in order, and find D_{snow} at the beginning of the month from

$$D_{snow} = D_{snow} \text{ in previous month} + \Delta D_{snow} \text{ in previous month}$$

7. Continue with step 2.

Discussion of Results

1. What is the maximum depth of snowpack? What portion of the total annual precipitation (depth ÷ annual precipitation) does this represent? In which month does it occur?

2. For each month, find the net precipitation as a proportion of "new" precipitation (P_{net} ÷ P). What are the highest and lowest values, and in which month does each occur?

3. If there were no snowpack, direct runoff would be highest in the month with highest P. Which month is this? With snowpack, direct runoff will be highest in the month with greatest net precipitation. How does the timing of maximum direct runoff change as a result of snowpack storage?

4. Based on the viewpoint given by the results of the previous questions, how important is it, in your judgment, to take snow storage into account in the water balance for your site?

DIRECT RUNOFF

Chapter 4 described estimation of direct runoff during individual storm events. This section describes the estimation of total monthly direct runoff, which results from the composite of storm and snowmelt events of all intensities and durations during each month.

Increases in direct runoff, with the resulting flooding, channel erosion, and loss of riparian habitat, are among the most crucial outcomes of urban development. Direct runoff also represents a loss to the long-term life of the landscape: it is subtracted directly from precipitation and immediately discharged from the site. Only the remaining moisture infiltrates the soil and participates in evapotranspiration, groundwater recharge, and stream base flow.

Ferguson (1996) provided a simple and valid method for estimating monthly direct runoff. It closely replicates the runoff that would be obtained by applying the SCS Method to daily precipitation each day during the month and summing the results. It is a simple formula:

$$Q_d = 0.208 \, P / S^{0.66} - 0.095$$

where

Q_d = average monthly direct runoff, inches

P = average monthly precipitation, inches; use the value of net precipitation P_{net} if you estimated snow storage

S = retention factor in the SCS Method; $S = 1,000/CN - 10$

Figure 5.5 is a chart that applies this formula. To use the chart, enter from the bottom with the monthly precipitation (or with net precipitation, if you calculated snow storage). Move vertically up to the line for the SCS curve number for your site, thence to the left to read the monthly direct runoff.

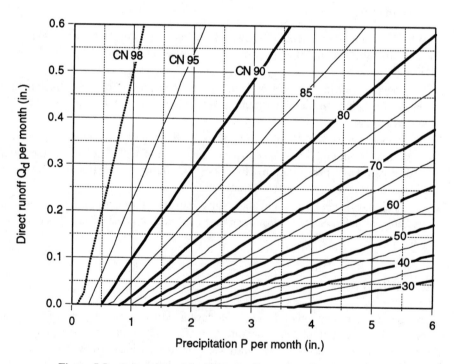

Figure 5.5 Average monthly direct runoff (based on Ferguson, 1996).

Direct Runoff Exercise

Exercise 5.2 finds direct runoff before and after development, so the results can be compared. Fill in the blanks with your data and calculations for the same sites you used in Exercises 4.1 through 4.4, and continue using the same sites for the rest of the exercises.

Summary of Process

1. Obtain the SCS curve number for your site. If you completed the storm runoff exercises in Chapter 4, then use the curve numbers you used in Exercise 4.3 (before development) and Exercise 4.4 (after development). If you did not complete the storm runoff exercises, then obtain curve number values from the tables and charts in Chapter 4.
2. Obtain monthly precipitation P in inches. If you calculated snowpack storage, use the value of monthly net precipitation P_{net}.
3. Find monthly direct runoff from Figure 5.5.

Discussion of Results

1. Before and after development, what is the proportion of total annual direct runoff to total annual precipitation $(Qd \div P)$? What is the highest proportion of monthly Qd to monthly P, and in which month does it occur? What is the lowest, and in which month does it occur? How does this proportion vary with increasing P?
2. How does this proportion compare with design storm runoff as a proportion of storm rainfall, found in Exercises 4.3 (before development) and 4.4 (after development)? Why is the proportion for monthly direct runoff different from that for design storm runoff?

POTENTIAL EVAPOTRANSPIRATION

Evapotranspiration is the composite of evaporation from water bodies and land surfaces, and transpiration through the leaves of plants. Evapotranspiration is the use of moisture by the on-site ecosystem for maintenance and development. Many people are astonished when they find out what a large portion of the annual rainfall ends up leaving their sites this way.

Evapotranspiration is produced by the interaction of a site's climate and soil. The climate produces a potential evapotranspiration by supplying heat and solar energy. Given that potential, the soil limits evapotranspiration if the amount of water in the soil is limited. Evapotranspiration Et can never exceed potential evapotranspiration Et_o, but it can be less. This section describes a climate's potential evapotranspiration Et_o; the following section compares Et_o with soil moisture to derive the actual evapotranspiration Et.

Thornthwaite's (1948) method of estimating Et_o is simple and valid. It uses easy-to-obtain input data of temperature and latitude. Other estimation methods have deeper theoretical grounding than Thornthwaite's, but they require more and harder-to-get input data without always producing more accurate estimates (Jensen, Burman, and Allen, 1990). In a few locales, such as California and central Arizona, potential evapotranspiration is monitored by government agencies; data for specific stations are available from extension service personnel.

Exercise 5.2 Direct runoff before and after development. All monthly values are in inches. Complete this exercise for both Site 1 and Site 2.

	Jan.	Feb.	Mar.	Apr.	May	June	July	Aug.	Sept.	Oct.	Nov.	Dec.
					Climatic data							
Precipitation *P* (from data source) =												
				Direct runoff before development								
SCS curve number before development (from Exercise 4.3) =												
Direct runoff Q_d (from Figure 5.5) =												
					Direct runoff after development							
SCS curve number after development (from Exercise 4.4) =												
Direct runoff Q_d (from Figure 5.5) =												

96

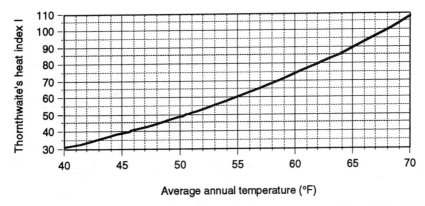

Figure 5.6 Heat index I in Thornthwaite's Et_o estimation method (formula $I = 9.189 + 0.0010285T^{2.7}$ was derived from data in Palmer and Havens, 1958).

Thornthwaite's method uses a site-specific heat index I to indicate the way Et_o varies with monthly temperature. Figure 5.6 shows the value of I as a function of a site's average annual temperature.

Potential evapotranspiration also varies with latitude, because day length and solar energy vary with latitude. Thornthwaite used the equator as a reference point for latitude: for a given value of I, each monthly temperature indicates equivalent Et_o at the equator; a later step converts the Et_o at the equator to that at the site. Figure 5.7 shows Et_o at the equator as a function of monthly temperature, for various values of I. To use the chart, enter from the bottom with the monthly average temperature. Move vertically up to the line for your site's heat index I. Thence move to the left to read the monthly Et_o that would occur at the equator.

Figure 5.8 shows the factor for converting Et_o at the equator to that at the latitude of your site. The factor varies by month, because the relationship of day length at your site to that at the equator changes with the seasons. To use the chart, enter from the bottom at the month you are calculating for. Move up to the line for the latitude of your site, thence to the left to read the ratio of Et_o at the site to that at the equator. Multiply the ratio by the Et_o at the equator to find the Et_o at the site. Repeat for each month of the year.

Potential Evapotranspiration Exercise

Exercise 5.3 finds the potential evapotranspiration Et_o for your site. Et_o is the same before and after development; it depends only on the climate.

Summary of Process

1. Obtain the latitude of your site from a map or other reference.
2. Obtain the monthly average temperature in degrees Fahrenheit. Find the average annual temperature (sum of monthly averages ÷ 12).
3. Find heat index I from Figure 5.6. The remaining steps are for each month separately.
4. Find evapotranspiration Et_o at the equator from Figure 5.7.
5. Find the ratio of Et_o at the site to Et_o at the equator from Figure 5.8.

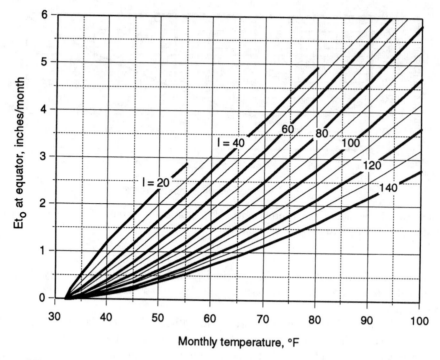

Figure 5.7 Et_o at the equator as a function of monthly temperature ("unadjusted potential evapo-transpiration" calculated by method of Thornthwaite, 1948).

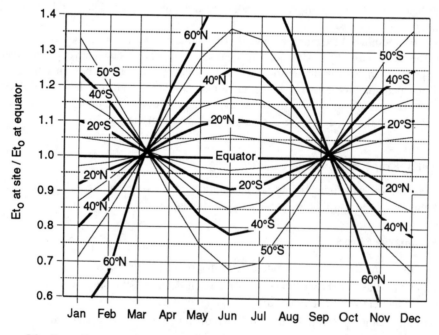

Figure 5.8 Factor for converting Et_o at the equator to Et_o at a site (based on data in Dunne and Leopold, 1978, p. 138).

EXERCISE 5.3   Potential evapotranspiration. Values of Et_o are in inches. Complete this exercise for both Site 1 and Site 2. Potential evapotranspiration is the same before and after development.

	Jan.	Feb.	Mar.	Apr.	May	June	July	Aug.	Sept.	Oct.	Nov.	Dec.
					Site data							
Latitude (°) (from map)	= _____											
					Climatic data							
Average monthly temperature (°F) (from data source)	= _____	____	____	____	____	____	____	____	____	____	____	____
Average annual temperature (°F) = sum of months ÷ 12	= _____											
Heat index I (from Figure 5.6)	= _____											
							Potential evapotranspiration Et_o					
Et_o at equator (from Figure 5.7)	= _____	____	____	____	____	____	____	____	____	____	____	____
Ratio Et_o at site / Et_o at equator (from Figure 5.8)	= _____	____	____	____	____	____	____	____	____	____	____	____
Et_o at site = ratio × Et_o at equator	= _____	____	____	____	____	____	____	____	____	____	____	____

99

6. Find Et_o at the site from

$$Et_o \text{ at site} = Et_o \text{ at equator} \times (Et_o \text{ at site} / Et_o \text{ at equator})$$

Discussion of Results

1. Which month has the highest Et_o? Which has the lowest? How does the amount in the highest month compare with the amount of precipitation in that month?
2. What is the total annual Et_o? How does this compare with total annual precipitation?

SOIL MOISTURE AND EVAPOTRANSPIRATION

As potential evapotranspiration "pulls" on the soil to extract the moisture, capillary tension in the soil resists. It resists more strongly as the soil dries out. The amount of moisture that is pulled out during a month represents a balance between potential evapotranspiration and the soil's resistance to it. Month by month, the stage is set for the creation of this balance by the soil's water-holding characteristics and the degree of drying it has already experienced.

Figure 5.9 shows the amount of water that different soils can hold in their root zones, available to be picked up by plant roots for transpiration. The available water-holding capacity (*AWHC*) varies with soil texture because of the soil's varying numbers and sizes of pore spaces. It also varies with type of vegetation, because the vegetation's rooting depth establishes the total volume of soil from which the roots can withdraw water. The assumed rooting depths of different types of vegetation are listed in Table 5.3. On a site that contains areas of impervious cover or different types of soil or vegetation, find the weighted average *AWHC* for the site as a whole.

Figure 5.10 shows the amount of water the soil retains when it is "pulled upon" by different levels of cumulative potential water loss. When sufficient moisture enters the soil to satisfy all potential evapotranspiration, the potential water loss is zero and the soil retains moisture equal to its full *AWHC*. When potential evapotranspiration exceeds the amount of moisture reaching the soil, then there is a potential water loss and the pulling begins. The potential water loss can accumulate from month to month in a dry season. The amount of moisture remaining in the soil, shown in Figure 5.10, reflects the balance between the opposing "pulls" of potential evapotranspiration and soil capillary tension. To use the chart, enter from the bottom at the cumulative potential water loss for the month. Move up vertically to the line for the soil's *AWHC*, thence to the left to read the amount of water remaining in the soil.

For each month, this balance is found through a series of steps that concludes with the soil moisture remaining in the soil and the resulting evapotranspiration. The steps are listed in the following "Summary of Process" for Exercise 5.4. These steps use the following terms for each month:

PMI = potential moisture increment, the potential inflow or outflow produced each month by the climate

CPG = cumulative potential gain, the cumulative sum of the positive values of *PMI*

CPL = cumulative potential loss, the cumulative sum of the negative values of *PMI*

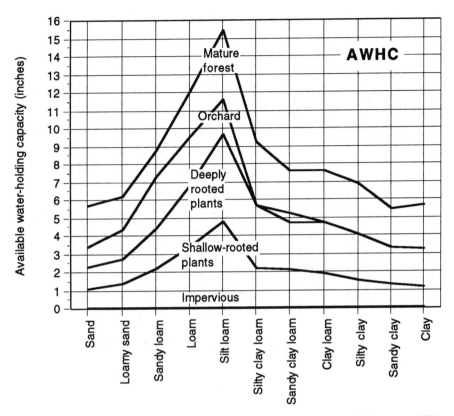

Figure 5.9   Soil available water holding capacity *AWHC* (based on data in Ferguson, 1994, pp. 108–109).

For sites with dry climates, a special kind of care must be taken in finding the cumulative potential loss *CPL*. On sites that are sufficiently dry, soil moisture may never rise to equal the soil's water-holding capacity. There is a cumulative potential loss even at the end of any wet season, because the soil is experiencing a water shortage year-round. For the last month of the wet season, you have to find the equivalent potential loss that is in equilibrium with the year-round net water shortage. This value depends on the ratio $\Sigma PL/AWHC$ (the sum of all the negative values of *PMI* divided by *AWHC*) and the ratio $\Sigma PG/AWHC$ (the sum of all the positive values of *PMI* divided by *AWHC*). The equilibrium *CPL* can be

TABLE 5.3   Typical characteristics of vegetation types referred to in Figure 5.9 (Thornthwaite and Mather, 1957, p. 244)

	Typical Root Zone Depth	Typical Examples
Shallow-rooted plants	1 to 2 ft	Spinach, peas, beans, beets, carrots
Deeply rooted plants	3 to 4 ft	Alfalfa, pastures, shrubs
Orchard	3 to 5 ft	Small trees
Mature forest	5 to 8 ft	Large trees, closed canopy

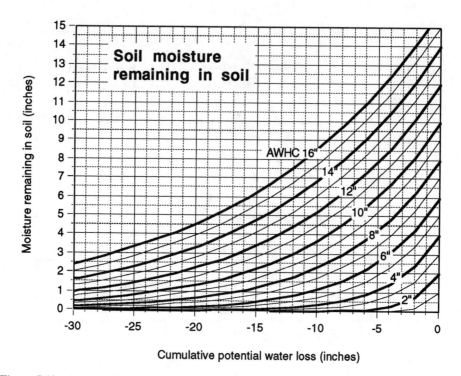

Figure 5.10 Amount of water remaining in the root zone as a function of *AWHC* and cumulative potential water loss (based on exponential loss equation).

found from the chart in Figure 5.11 by entering from the bottom with the site's $\Sigma PL/AWHC$, moving vertically up to the line for the site's $\Sigma PG/AWHC$, and thence to the left to read the equilibrium ratio $CPL/AWHC$. Apply this ratio to find the equilibrium CPL at the end of the wet season,

$$\text{Equilibrium } CPL = AWHC \times (CPL/AWHC)$$

Assign this amount to the last month of the wet season, as was done for March in the Los Angeles calculations in Table 5.2. Add it to the monthly CPL in the first month of the dry season, to begin figuring cumulative potential loss when the dry season actually begins. Complete the rest of the calculations normally.

Soil Moisture and Evapotranspiration Exercise

Exercise 5.4 finds soil moisture and evapotranspiration before and after development, so the results can be compared.

Summary of Process

1. Obtain the soil's available water-holding capacity *AWHC* from Figure 5.9. Obtain monthly precipitation *P* and direct runoff *Qd* from Exercise 5.2, and potential evapotranspiration Et_o from Exercise 5.3. Carry out the following steps for each month separately.

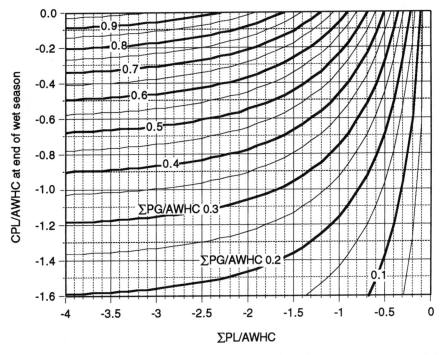

Figure 5.11 Equilibrium *CPL/AWHC* at the end of the wet season, in dry climates (derived from algorithm described in Thornthwaite and Mather, 1957, p. 190).

2. Find the potential moisture increment *PMI*:

$$PMI = P - Q_d - Et_o$$

3. Find the cumulative potential gain CPG:

If $PMI > 0$: $CPG = PMI +$ previous month's *CPG*

If $PMI \leq 0$: $CPG = 0$

4. Find the cumulative potential loss *CPL*:

If $PMI > 0$: $CPL = 0$

If $PMI \leq 0$: $CPL = PMI +$ previous month's *CPL*

5. Find soil moisture storage *SM*:

If $PMI > 0$: $SM = PMI +$ previous month's *SM*
            (up to the soil's *AWHC*)

If $PMI \leq 0$:   find *SM* from Figure 5.10

6. Find change in soil moisture Δ*SM*:

$$\Delta SM = SM - \text{previous month's } SM$$

Exercise 5.4 Soil moisture and evapotranspiration before and after development. All values are in inches. Obtain P, Q_d and Et_o from Exercises 5.2 and 5.3. Complete this exercise for both Site 1 and Site 2.

	Jan.	Feb.	Mar.	Apr.	May	June	July	Aug.	Sept.	Oct.	Nov.	Dec.

Soil moisture and evapotranspiration before development

Soil $AWHC$ before development
(from Figure 5.9) = _____

Potential moisture increment PMI
$= P - Q_d - Et_o$ =

Cumulative potential gain CPG (for $PMI > 0$ only)
$= PMI + $ previous CPG =

Cumulative potential loss CPL (for $PMI \leq 0$ only)
$= PMI + $ previous CPL =

Soil moisture storage $SM = PMI + $ previous SM,
or from Figure 5.10 =

Change in soil moisture ΔSM
$= SM - $ previous SM =

Evapotranspiration Et
$= Et_o$ or $P - Q_d - \Delta SM$ =

Soil moisture and evapotranspiration after development

Soil $AWHC$ after development
(from Figure 5.9) = _____

Potential moisture increment PMI
$= P - Q_d - Et_o$ =

Cumulative potential gain CPG (for $PMI > 0$ only)
$= PMI + $ previous CPG =

Cumulative potential loss CPL (for $PMI \leq 0$ only)
$= PMI + $ previous CPL =

Soil moisture storage $SM = PMI + $ previous SM,
or from Figure 5.10 =

Change in soil moisture ΔSM
$= SM - $ previous SM =

Evapotranspiration Et
$= Et_o$, or $P - Q_d - \Delta SM$ =

7. Find evapotranspiration *Et:*

$$\text{If } PMI > 0: \qquad Et = Et_o$$

$$\text{If } PMI \leq 0: \qquad Et = P - Qd - \Delta SM$$

Discussion of Results

1. What is the highest level of soil moisture storage *SM* in any one month? What is the lowest? In which month does each occur? Why does the soil moisture regime show this seasonal pattern? How does vegetation in your locale respond to this seasonal moisture environment?

2. What is the total annual evapotranspiration? How does it compare with total annual potential evapotranspiration? With total annual precipitation?

3. Which month has the greatest excess of precipitation over evapotranspiration (if any)? Which month has the greatest deficit of evapotranspiration below potential evapotranspiration? How do stream flow, lake levels, and other natural processes in your locale respond to this seasonal pattern?

4. Does any month experience evapotranspiration greater than precipitation? If so, how is this possible? Where did the water for evapotranspiration come from?

GROUNDWATER AND BASE FLOW

Excess soil moisture drains into the groundwater, where it sustains base flows.

In months when soil moisture *SM* is at the soil's water holding capacity, any additional precipitation is a surplus that adds to the groundwater. The existence of a surplus is indicated by a positive change in soil moisture ΔSM. In a month with a surplus, the amount added to groundwater, the groundwater increment *GWI*, is equal to $PMI - \Delta SM$.

The total amount in groundwater storage is equal to the month's *GWI* plus the amount of groundwater detained from the previous month. Of the amount in storage, it is assumed that half drains out as base flow during the month. The other half remains in storage, to be made available for discharge in the next month.

Finding the amounts of groundwater and base flow concludes the water balance. For each month, all the site's inflows, outflows, and changes in storage can be summarized:

$$P = Et + Qd + \text{base flow} + \Delta\text{snow} + \Delta SM + \Delta\text{groundwater}$$

The equation says that inflowing precipitation is partitioned into outflows of *Et*, *Qd*, and base flow; any imbalance between precipitation and outflow comes from a change in storage in some combination of snowpack, soil moisture, and groundwater. The seasonally changing amounts in these storage compartments reflect an equilibrium between the site's precipitation, temperature, vegetation, and soil. Each site has its own equilibrium and its own susceptibility to change as a result of urban development.

Over the course of an average year, the change in storage of all kinds is zero. The water balance reduces to the year's total inflows and outflows:

$$P = Et + Qd + \text{base flow}$$

All the calculations previously described are in units of depth (inches) of water. To find the volume of any flow or storage, either per month or per year, multiply depth by the area of the site A_d, and use 12 as the conversion factor from inches to feet:

$$\text{Volume (af)} = \text{depth (in.)} \times A_d \text{ (ac)} / 12$$

Groundwater and Base Flow Exercise

Exercise 5.5 finds groundwater and base flow before and after development, so the results can be compared.

Summary of Process

1. Obtain each month's potential moisture increment *PMI* and change in soil moisture ΔSM before or after development from Exercise 5.4.
2. Find groundwater increment *GWI*:

$$\text{If } \Delta\text{SM} < 0: \quad GWI = 0$$

$$\text{If } \Delta\text{SM} \geq 0: \quad GWI = PMI - \Delta\text{SM}$$

3. Find groundwater storage *GW*:

$$GW = GWI + \text{previous month's groundwater detention}$$

4. Find base flow:

$$\text{Base flow} = 0.5 \ GW$$

5. Find groundwater detention:

$$\text{Groundwater detention} = 0.5 \ GW$$

6. Find total stream flow $q\Sigma$:

$$q\Sigma = Q_d + \text{base flow}$$

Discussion of Results

1. What is the lowest monthly amount of base flow? In which month does it happen? Why? How does low base flow affect aquatic life, municipal water supply, and the susceptibility of streams to pollution?
2. What is the total volume of precipitation entering your site in an average year, in acre feet per year? In what ways is this quantity of water a resource for people and the ecosystem? In what ways is this resource renewable, or not renewable?

SUMMARY AND COMMENTARY

The sustainability of natural and human communities is as much a result of long-term flows and their fluctuation from season to season, as of isolated design storms. The support

Exercise 5.5 Groundwater and base flow before and after development. All values are in inches. Obtain *PMI* and ΔSM from Exercise 5.4. Complete this exercise for both Site 1 and Site 2.

	Jan.	Feb.	Mar.	Apr.	May	June	July	Aug.	Sept.	Oct.	Nov.	Dec.

Groundwater and base flow before development

Groundwater increment *GWI* (for $\Delta SM \geq 0$ only)
= *PMI* - ΔSM = ___ ___ ___ ___

Groundwater storage *GW*
= *GWI* + previous detention = ___ ___ ___ ___

Base flow
= 0.5 *GW* = ___ ___ ___ ___

Groundwater detention
= 0.5 *GW* = ___ ___ ___ ___

Total stream flow *q*
= Q_d + base flow = ___ ___ ___ ___

Groundwater and base flow after development

Groundwater gain increment (for $\Delta SM \geq 0$ only)
= *PMI* − ΔSM = ___ ___ ___ ___

Groundwater storage *GW*
= *GWI* + previous detention = ___ ___ ___ ___

Base flow
= 0.5 *GW* = ___ ___ ___ ___

Groundwater detention
= 0.5 *GW* = ___ ___ ___ ___

Total stream flow *q*Σ
= Q_d + base flow = ___ ___ ___ ___

of perennially flowing streams and the recharge of underlying aquifers depend on the cumulative effect of numerous small storms, their storage in the soil, and their relationship to evapotranspiration. The water balance is a way to evaluate the aggregate effect of the hydrologic regime.

REFERENCES

Dunne, Thomas, and Luna B. Leopold, 1978, *Water in Environmental Planning*, San Francisco: Freeman.

Ferguson, Bruce K., 1994, *Stormwater Infiltration*, Boca Raton: Lewis.

Ferguson, Bruce K., 1996, Estimation of Direct Runoff in the Thornthwaite Water Balance, *Professional Geographer* vol. 48, no. 3, pp. 263–271.

Ferguson, Bruce K., M. Morgan Ellington, and P. Rexford Gonnsen, 1991, Evaluation and Control of the Long-Term Water Balance of an Urban Development Site, in *Proceedings of the 1991 Georgia Water Resources Conference*, Kathryn Hatcher, editor, pp. 217–220, Athens: University of Georgia Institute of Natural Resources.

Jensen, M. E., R. D. Burman, and R. G. Allen, editors, 1990, *Evapotranspiration and Irrigation Water Requirements*, Manuals and Reports on Engineering Practice No. 70, New York: American Society of Civil Engineers.

Mather, John R., 1978, *The Climatic Water Balance in Environmental Analysis*, Lexington: Lexington Books.

Palmer, Wayne C., and A. Vaughn Havens, 1958, A Graphical Technique for Determining Evapotranspiration by the Thornthwaite Method, *Monthly Weather Review* vol. 86, pp. 123–128.

Thornthwaite, C. W., 1948, An Approach Toward a Rational Classification of Climate, *Geographical Review* vol. 38, no. 1, pp. 55–94.

Thornthwaite, C. W, and J. R. Mather, 1955, *The Water Balance*, Centerton, N.J.: Drexel Institute of Technology, Publications in Climatology vol. 8, no. 1.

Thornthwaite, C. W, and J. R. Mather, 1957, *Instructions and Tables for Computing Potential Evapotranspiration and the Water Balance*, Centerton, N.J.: Drexel Institute of Technology, Publications in Climatology vol. 10, no. 3.

CHAPTER 6

CONVEYANCE

Conveyance is the moving of water across a land surface. It is the most ancient of stormwater management functions. In the 2,000-year-old streets of Pompeii you can see systems of gutters that drain the city's stormwater out to the rivers and the sea.

Conveyance systems are characterized by linear pipes and swales draining one into another. In plan view each conveyance connects to adjacent ones. In profile view the invert elevation—the bottom elevation at which water flows—slopes continuously downward through pipes and swales alike (Figure 6.1). All the water flows across the land surface; the configuration of the system controls the rate.

Until about 1965 conveyance was the exclusive stormwater approach in almost all urban areas of the United States. It got rid of on-site nuisances. That is still one of its purposes, and you can still see conveyance systems being built everywhere. In parts of many sites conveyance is, in fact, the only reasonable way to treat stormwater and alternative measures are not worth the effort. But today conveyance, in conjunction with other approaches described in later chapters of this book, is also one of the steps in managing flow to treat runoff and restore the natural hydrologic balance.

A pipe or swale must be capable of conveying the peak flow rate during a design storm. If it is big enough to carry that flow, then it can also handle all the rest of the flow during the design storm, as well as low flows at other times. Larger sizes, especially of pipes, would cost more, so in making a conveyance exactly the size necessary for design-storm capacity

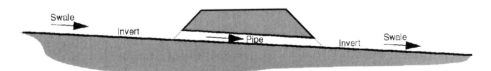

Figure 6.1 Continuity of the invert elevation in conveyance.

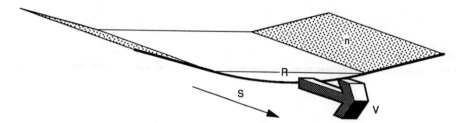

Figure 6.2   Flow characteristics related by Manning's equation.

you are choosing a balance between cost and frequency of overflow. Larger conveyance sizes may be selected for aesthetic, environmental, or safety reasons when cost allows.

When a storm larger than your design storm occurs, your conveyance overflows. Although you accepted this risk in selecting a particular design storm, you still have to provide, by appropriate grading, an emergency overflow—a safe, nonerodible, secondary drainage system for excess water to follow when a larger storm occurs or the primary system is clogged.

Different configurations of swales and pipes control rate of flow in different ways. This chapter describes the flow controls in four types of conveyances: swale, weir, orifice, and pipe. But first you need Manning's equation, which is a general tool for the analysis of flow conditions in any conveyance.

MANNING'S EQUATION

Manning's equation describes uniform flow. Uniform flow occurs in a conveyance to which runoff is delivered at a steady rate and where the size, shape, gradient, and direction of the conveyance are uniform along its length.

Manning was an Irishman who did most of his work late in the nineteenth century. In a laboratory today you can recreate appropriate conditions and derive essentially the same relationship he did. Despite the age of Manning's equation it is still valid and useful—as long as you apply it only where the flow is uniform. Figure 6.2 shows the flow characteristics that are related by the equation.

Manning's equation is:

$$V = (1.49 \, / \, n) \, R^{2/3} \, G^{1/2}$$

where

V = velocity (fps)

n = roughness factor (no units); values are given in Table 6.1

R = hydraulic radius; $R = A/W_p$, where, as shown in Figure 6.3

   A = cross-sectional area (sf) through which water is flowing

   W_p = wetted perimeter (ft), the cross-sectional length of surface in contact with water

G = slope along conveyance's length (ft/ft)

TABLE 6.1   Values of roughness factor *n* in Manning's equation (from Table 2 of U.S. Federal Highway Administration, 1973b, and Table B-6, p. 577, of U.S. Bureau of Reclamation, 1974, except where shown)

Conveyance Material	n
Concrete pipe	0.013
Corrugated metal pipe	0.024
Brick	0.014–0.017
Concrete swale, trowel finish	0.012–0.014
Random stone in mortar	0.020–0.023
Dry rubble (riprap) (Chow, 1959, p. 111)	0.033
Dry rubble (riprap) (Table 2 in Abt et al., 1988):	
1 in.–1.25 in. crushed stone on 1–2% slope	0.024
1 in.–1.25 in. crushed stone on 10% slope	0.055
2.2 in.–2.8 in. crushed stone on 2% slope	0.025
2.2 in.–2.8 in. crushed stone on 8–10% slope	0.030
Asphalt	0.013–0.016
Earth with short grass, few weeds (turf)	0.022–0.027
Earth with dense weeds and high brush	0.08–0.12
Earth, clean bottom, brush on sides	0.05–0.08
Sheet flow; consider cover to height of 0.1 ft only (SCS 1986, Table 3.1):	
Concrete, asphalt, gravel, or bare soil	0.011
Dense grasses such as weeping lovegrass, bluegrass, buffalo grass, blue grama grass, and native grass mixtures	0.24
Bermuda grass	0.41
Natural range	0.13
Woods with light underbrush	0.40
Woods with dense underbrush	0.80
Large natural stream channels (adapted from Barnes, 1967):	
Earthen banks with some brush	0.026
Banks covered with low reeds and grass	0.027
Banks covered with low weeds and brush	0.030
Banks and bed of uniform smooth stones	0.032
Banks and bed of smooth stones; some debris and brush	0.037
Banks thickly covered with woody brush and trees	0.040–0.060
Angular boulders in channel bed	0.041–0.050
Bed mostly of boulders that deflect flow	0.055–0.075
Small natural stream channels (Table 16.1 in Dunne and Leopold, 1978):	
Winding channels considerably covered with small growth	0.035
Streams with bank or aquatic vegetation	0.040–0.050
Mountain streams in clean loose cobbles	0.040–0.050
Irregular alignment and cross section, obstructed by trees and brush	0.100
Very irregular alignment and cross section, many roots, trees, logs, drift on bottom	0.150–0.200
Densely wooded floodplain (generalized from Arcement and Schneider, 1989):	
Negligible undergrowth, all trees large, soil surface smooth	0.10
Negligible undergrowth, mix of large and small trees	0.12–0.15
Dense undergrowth	0.20

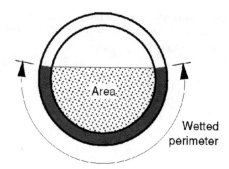

Figure 6.3   Area and wetted perimeter in a conveyance.

Roughness n indicates how much a material resists flow. The values in Table 6.1 range from 0.012 to 0.8, so the choice of conveyance material and vegetation, and thus of n, alone could have more than a 60-fold effect on a conveyance's velocity and capacity. For precise roughness values of natural channels, the publications of Barnes (1967) and Arcement and Schneider (1989) provide photographs and detailed tables that are very helpful.

Hydraulic radius R indicates how much cross-sectional area is available for carrying water for a given length of frictional contact with the conveyance's sides. A broad, shallow swale has a small cross-sectional area in proportion to the wetted perimeter and tends relatively to resist flow. In contrast, a circular pipe, when flowing mostly full, has a large hydraulic radius and tends to have high velocity and high conveyance capacity in relation to its size.

Slope G indicates how directly gravity can pull water through a conveyance. As slope increases, velocity increases.

Manning's equation can be solved for rate of flow q. Because $V = q/A$ you can substitute for V in the original equation and derive, with q in cfs,

$$q = (1.49 / n) \, A \, R^{2/3} \, G^{1/2}$$

Figure 6.4 shows an application of the Manning equation to a conveyance problem. The stream is in Charlotte, North Carolina. Just upstream from this view, a public housing project became subject to stream flooding following urbanization of the watershed. Straightening and smoothing the banks lowers Manning's roughness to half its preexisting value. In consequence, the capacity of the channel will double and flooding will be reduced. In making the decision to disturb the stream this way, the planners had to weigh the habitat value of the ecologically complex stream banks against the safety of the nearby residents. Upstream detention to reduce peak flows was investigated as an alternative to stream straightening, but was not considered feasible because all the land in the watershed was already occupied by development.

Manning's equation is complicated enough that practitioners seldom try to solve it by hand. This chapter supplies charts that solve the equation for common applications. Use of a computer spreadsheet set for iterative calculation is another way to implement the equation.

Analytical Exercises

1. Using algebra, derive from Manning's equation a special-purpose equation for velocity in a circular pipe flowing full. Replace hydraulic radius with terms for the pipe's

Figure 6.4   Straightening and smoothing of a creek channel in Charlotte, North Carolina.

circumference and area. Then simplify the equation as much as possible so that it relates velocity to, among other things, the pipe's diameter.

2. Using algebra, derive from Manning's equation a special-purpose equation for velocity in shallow sheet flow across a land surface. Simplify the equation as much as possible.

3. Where a pipe or channel already exists or has been designed, and you know the rate of flow q entering it, how can you use Manning's equation to find time of travel through the conveyance?

SWALES

Swales are open channels with unobstructed flow. They are vital parts of almost any drainage system and any landscape. Drainage in contact with soil increases vegetative vari-

TABLE 6.2 Maximum noneroding velocities in dense, uniform grass cover (from Tables 3 and 4 in U.S. Federal Highway Administration, 1973b)

	On Erosion-Resistant Soil (stiff clay, silt, hardpan, fine and coarse gravel)			On Easily Eroded Soil (fine sand, sandy loam, loam, silt and clay loam)		
	Channel Slope 0–5%	Channel Slope 5–10%	Channel Slope 10%+	Channel Slope 0–5%	Channel Slope 5–10%	Channel Slope 10%+
Bermuda grass	8 fps	7 fps	6 fps	6 fps	5 fps	4 fps
Buffalo grass, Kentucky bluegrass, smooth brome, blue grama	7 fps	6 fps	5 fps	5 fps	4 fps	3 fps
Grass mixture	5 fps	4 fps	erodible	4 fps	3 fps	erodible
Lespedeza sericea, weeping lovegrass, yellow bluestem, kudzu, crabgrass, common lespedeza, Sudan grass	3.5 fps	erodible	erodible	2.5 fps	erodible	erodible

ety, reduces velocity, decreases downstream peak flow, permits infiltration, symbolizes interaction with nature, and supports wildlife habitat and potential human amenity (Thayer and Westbrook, 1990). Flow in a swale tends to be uniform, so it can usually be evaluated with Manning's formula.

The allowable width or depth of flow in a swale may be limited by the locations and elevations of nearby buildings, roads, and utility lines. You can identify such limitations by examining a topographic map of the site or a profile of the drainage system. To limit flow width or depth, you could change the swale's cross-sectional shape. Alternatively, you could increase velocity by reducing the channel's roughness or increasing its slope; according to $q=VA$ this would reduce the cross-sectional area of flow. You could limit q_p that the swale has to carry by installing detention or infiltration basins upstream or by revising the site plan to divert runoff to other discharge points. You could drop the flow directly to a lower elevation through a drop structure.

The allowable velocity in a swale may by limited by the need to prevent erosion. A swale ought to be stable during the storm for which it is designed. Maximum noneroding velocities for different channel materials and covers are listed in Tables 6.2 through 6.4. Erosive

TABLE 6.3 Maximum noneroding velocities in crushed stone (from Table 12.10, pp. 12–90 in Moulton, 1991; and National Stone Association, 1978)

National Stone Association Designation	Minimum Size	Average Size	Maximum Size	Minimum Thickness of Stone Layer	Velocity
R-2	1.0 in.	1.5 in.	3.0 in.	3.0 in.	4.5 fps
R-3	2.0 in.	3.0 in.	6.0 in.	6.0 in.	6.5 fps
R-4	3.0 in.	6.0 in.	12.0 in.	12.0 in.	9.0 fps
R-5	5.0 in.	9.0 in.	18.0 in.	18.0 in.	11.5 fps
R-6	7.0 in.	12.0 in.	24.0 in.	24.0 in.	13.0 fps
R-7	12.0 in.	15.0 in.	30.0 in.	30.0 in.	14.5 fps

TABLE 6.4 Maximum noneroding velocities in natural materials without vegetation (from Chow, 1959, p. 165)

Material	Maximum Noneroding Velocity, fps
Fine sand	1.5
Fine gravel	2.5
Coarse gravel	4.0
Cobbles and shingles	5.0

velocity can be recognized in design by estimating design storm velocity and comparing the result with the maximum noneroding velocity for the swale's slope, soil, and vegetation type. Velocity can be found using Manning's equation or the equation $V = q/A$.

You can reduce velocity by reducing gradient with grade control structures, by increasing roughness with boulders, dense vegetation, or complex bank forms, or by revising the site plan to reduce the q_p that enters the swale. Swales with low velocity are characteristically densely vegetated, gently sloping, and broad. Reducing velocity tends to suppress downstream flood peaks. Reducing velocity increases flow height in the swale, so it may require adjacent floodway open space.

You can increase a channel's resistance to erosion with appropriate channel armoring (lining). In Albuquerque, New Mexico, where Arroyo Pino flows through an office park, a tree-lined jogging path meanders at the top of a grassed floodway embankment overlooking the concrete-lined low-flow channel. Concrete linings are very nonerodible and give high capacity to swales because of their low n values, but they obstruct soil infiltration and fail to provide biotic habitat. In contrast, riprap and gabions make porous, flexible linings that interact to a degree with the aquatic habitat.

Riparian vegetation can stabilize channel banks where the water's erosive power is not excessive. During high flows, resilient vegetation bends into erosion-inhibiting mats. Binding networks of roots increase soil shear strength. When vegetation is used together with nonliving materials, the approach is termed *bioengineering*. A vegetatively lined channel is the San Diego River in California, where a variety of species were planted to support riparian wildlife while stabilizing the steep banks of the floodway.

Triangular Swale

Figure 6.5 shows one kind of swale cross section. This is a triangular cross section with 10:1 (10 percent) side slopes. Such a broad, gently sloping cross section might be used in large open areas such as golf courses.

Figures 6.6 and 6.7 relate the velocity, depth, and slope in such a swale according to Manning's equation. Enter the appropriate chart at the bottom with the slope of your swale.

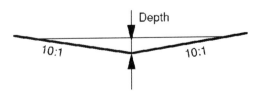

Figure 6.5 Cross section through a triangular swale.

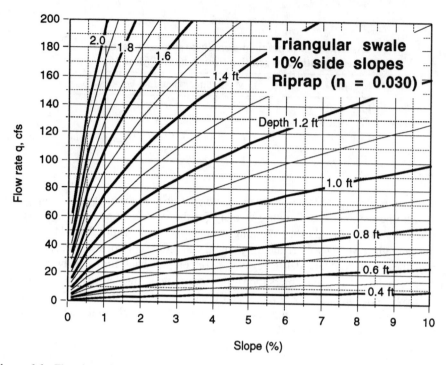

Figure 6.6 Flow in a triangular swale with 10 percent side slopes and riprap cover (data derived from Manning's equation).

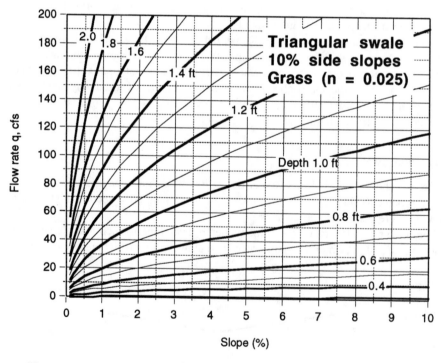

Figure 6.7 Flow in a triangular swale with 10 percent side slopes and turf cover (data derived from Manning's equation).

116

Then enter the chart again from the left side, with the flow rate. Find the point on the graph where your slope and flow rate intersect. Then interpolate between curves to find the depth of flow D.

Based on the geometry of the cross section, width of flow W is given by

$$W = 20\,D$$

Based on the equation $V = q/A$ and the cross section's geometry, velocity is given by

$$V = q\,/\,(10\,D^2)$$

where

$$V = \text{velocity, fps}$$

$$q = \text{rate of flow, cfs}$$

$$D = \text{depth at deepest (center) point of swale, ft}$$

Analytical Exercises

1. Using algebra, reproduce the derivation of the equation $V = q/(10\,D^2)$ from $q = V/A$. Begin by writing down a few equations that state what you can tell about the geometry of the triangular cross section. Then substitute in terms for $q = V/A$ and solve for velocity.
2. Derive another special-purpose velocity equation for this swale, beginning from Manning's equation.
3. Under what circumstances would you choose to use each of the two special-purpose velocity equations?
4. To what size of riprap does the chart in Figure 6.6 apply?

Swale Exercise

Exercise 6.1 contrasts grass swale designs based on flow rates derived by the Rational and SCS methods, and evaluates them according to velocity. Use the same sites that you used in Exercises 4.1 through 4.4, and continue using them for the rest of the exercises. The results of this exercise are used in Exercise 6.2.

Summary of Process

1. Obtain roughness factor n from Table 6.1.
2. From a topographic map, find the swale's slope and any site-specific limitations on allowable flow depth or width.
3. Obtain maximum noneroding velocity from Table 6.2.
4. Obtain q_p after development from the Rational storm runoff estimate in Exercise 4.2, and the SCS estimate in Exercise 4.4.

Exercise 6.1   Triangular swale, 10:1 side slopes, turf (for cross section, see Figure 6.5)

	Site 1	Site 2

Determining data

Manning's roughness n
   (from Table 6.1)   = _____   _____

Gradient of swale
   (from site plan)   = _____ %   _____ %

Maximum allowable width of flow
   (from site plan)   = _____ ft   _____ ft

Maximum noneroding velocity
   (from Table 6.2)   = _____ fps   _____ fps

Design based on Rational formula

q_p by Rational formula, after
   development (from Exercise 4.2)   = _____ cfs   _____ cfs

Flow depth D
   (from Figure 6.7)   = _____ ft   _____ ft

Width of flow
   = 20 D = _____ ft   _____ ft

Velocity
   = $q_p / (10\,D^2)$ = _____ fps   _____ fps

Design based on SCS method

q_p by SCS method, after development
   (from Exercise 4.4)   = _____ cfs   _____ cfs

Flow depth D
   (Figure 6.7)   = _____ ft   _____ ft

Width of flow
   = 20 D = _____ ft   _____ ft

Velocity
   = $q_p / (10\,D^2)$ = _____ fps   _____ fps

5. For each q_p estimate, find the swale's flow depth, using Manning's equation or Figure 6.7. Compute flow width W from $W = 20\,D$.

6. Compute velocity using Manning's equation or $V = q_p\,p / (10\,D^2)$.

7. If velocity exceeds the maximum allowable velocity, select a cross section with larger capacity, armor the erodible soil with a stable material, or revise the site plan to reduce swale gradient or rate of runoff reaching the swale. Then begin again with step 1.

Discussion of Results

1. Which site requires the deeper and wider swale? Why?

2. Which runoff estimation method requires the larger swale? Why? Wide swales pre-empt valuable urban land. What conveyance sizing procedure would you use in practice? Why?

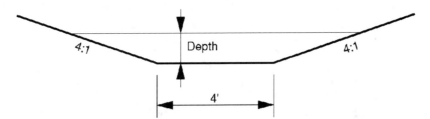

Figure 6.8 A trapezoidal swale with 4-ft bottom and 4:1 (25 percent) side slopes.

3. Is peak rate of flow in cfs the only consideration that may limit the size of a conveyance? Why? In what specific types of site conditions are other considerations most likely to limit conveyance size?

4. What are some approaches you might take to reducing velocity in a swale? If reducing the velocity is impractical, what are some other approaches you could take to reducing swale erosion?

Other Swale Shapes

Figure 6.8 shows another common cross-sectional swale shape, a trapezoid. Manning flow charts for this shape are shown in Figures 6.9 through 6.12.

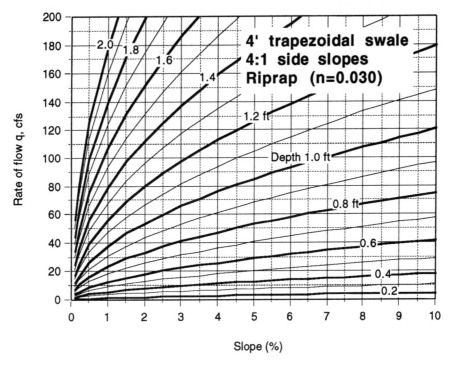

Figure 6.9 Flow in a trapezoidal swale with 4-ft bottom and 4:1 side slopes, riprap cover (data derived from Manning's equation).

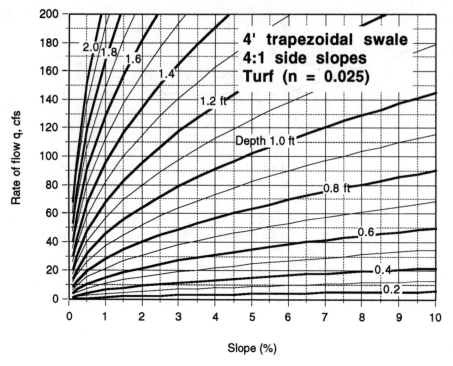

Figure 6.10 Flow in a trapezoidal swale with 4-ft bottom and 4:1 side slopes, turf cover (data derived from Manning's equation).

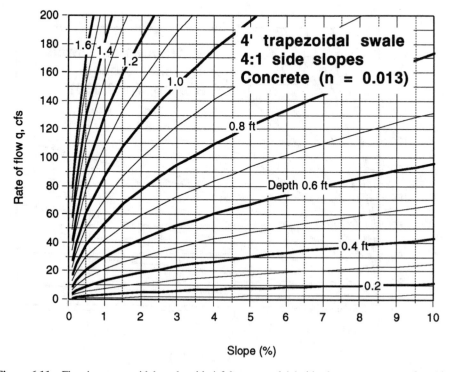

Figure 6.11 Flow in a trapezoidal swale with 4-ft bottom and 4:1 side slopes, concrete surface (data derived from Manning's equation).

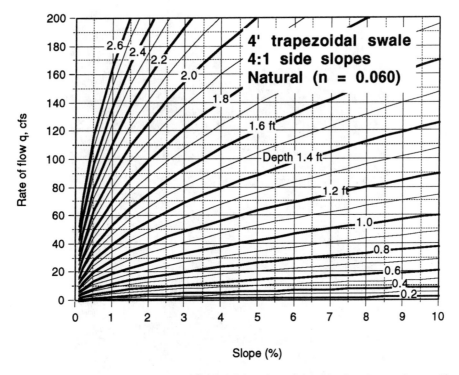

Figure 6.12 Flow in a trapezoidal swale with 4-ft bottom and 4:1 side slopes, natural cover (data derived from Manning's equation).

Figure 6.13 shows a swale of this shape under construction in Georgia. The purpose of this swale is to divert runoff from a previously developed site on one side of the swale, into a detention basin for control. The swale is very straight, and its dimensions are quite constrained because it had to fit in a narrow corridor between the developed site and a sanitary sewer.

According to $V = q/A$ and the geometry of the cross section, velocity in a swale of this shape is given by

$$V = q / (4D + 4 D^2)$$

where

$$V = \text{velocity in fps}$$

$$q = \text{rate of flow in cfs}$$

$$D = \text{depth in center of swale, in feet}$$

Other geometric swale shapes are possible, such as parabolic. Many floor widths and side slope gradients are possible. The publication *Design Charts for Open-Channel Flow* (Hydraulic Design Series No. 3) by the U.S. Federal Highway Administration (1973a) provides Manning flow charts for a wide variety of shapes and sizes of swales. Implementing

Figure 6.13   A trapezoidal swale under construction in Athens, Georgia.

Manning's equation in a computer spreadsheet is always an available way to derive a design using any swale shape.

Analytical Exercises

1. Using algebra, reproduce the derivation of the trapezoidal swale velocity equation from $V = q/A$. Begin by writing down a few equations that state what you can tell about the geometry of the trapezoidal cross section. Then substitute in terms for $V = q/A$ and solve for velocity.

2. Derive another special-purpose velocity equation for this swale, beginning from Manning's equation.

3. Under what circumstances would you choose to use each of the two special-purpose velocity equations?

Design to Improve Water Quality

As water flows through vegetated swales, its quality is affected to some degree. Unlike smooth, impervious gutters, pipes, and channels, vegetated swales store runoff while conveying it at low velocity and provide a large surface area in contact with vegetation and soil for biophysical treatment and infiltration.

Swales designed specifically for pollution reduction have been called *biofilters*, reflecting the use of that term in sewage treatment by shallow flow across a vegetated surface. The special criterion for runoff biofiltration is limiting velocity to 0.5 fps (Municipality of Metropolitan Seattle, 1992, pp. 1–3). Swales with this velocity capture 63 to 83 percent of particulate pollutants and those other pollutants that adhere to vegetation, including sediment, metals adsorbed on sediment, and oils. Swales are less effective for dissolved metals and nutrients (29 to 46 percent removal), and their effectiveness for bacteria is variable. Where a biofilter swale is added specifically for pollution control, it should be at least 200 feet long so that residence time in the swale is at least nine minutes.

Where low velocity cannot be maintained because of steep slope or constricted width, treatment performance can be reclaimed by ponding with check dams of stone, earth, or wood. Simple stone check dams are specified in many states' erosion and sediment control manuals and can cost less than $100 each. It is essential to space check dams closely enough to produce continuous ponding along the swale.

Capturing runoff in vegetated swales at all the sources of runoff in a development site uses the full capacity of the site's vegetation and soil for treatment and infiltration. It reduces the need for downstream reservoirs, which would disturb preexisting riparian zones and cost tens of thousands of dollars. Implementing this idea completely requires you to pay attention to the details of the site. Swales and check dams should be small, numerous, and located at all the runoff sources, including every roof, every downspout, every curb cut, and every small area of pavement.

Along streets, swales can replace curbs and gutters for drainage while adding the functions of runoff attenuation, infiltration, and treatment. Where curbs and gutters are appropriately applied, they protect the pavement edge structurally from overrunning traffic and they help organize on-street parking. However, they collect and concentrate pollutants while preventing runoff from being treated in contact with soil and vegetation. Therefore curbs should be installed only where they are actually needed for a specific street. Where curbs are in fact required, they can include cuts or notches to empty runoff at frequent intervals into vegetated swales. Every driveway across a swale can be used as a check dam (Figure 6.14).

In the Village Homes community in Davis, California, swales were graded to conform to the open space plan, reclaiming a site that had previously been smoothly graded for flood irrigation (Corbett, 1981; Thayer and Westbrook, 1990). Runoff drains into the swales through curb cuts and across sloping yards (Figure 6.15). Swales are landscaped like seasonal streambeds, with rocks, shrubs, and trees. Small wooden bridges span the swales for foot and bicycle traffic. Although the swale gradients and velocities are naturally low, wooden and rock check dams were added at places wide enough for ponding. Excess runoff spills over the small weirs and flows onto further series of swales and pools. Frogs, toads, newts, blackbirds, flickers, owls, doves, killdeers, opossums, skunks, raccoons, and numerous butterflies and moths have been identified in the drainage corridor.

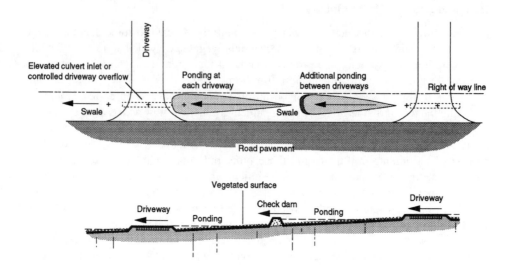

Figure 6.14   Ponding along a street swale.

Figure 6.15   Vegetated swales between the homes and walks of Village Homes in Davis, California.

Analytical Exercises

1. For the swale conditions used in Exercise 6.1, if velocity were reduced to 0.5 fps, what would the depth and width of flow be? Is there any swale material sufficiently rough that it might reduce velocity to this degree?

2. For the same swale conditions, what is the maximum spacing of 1-ft high check dams along the length of the swale that would produce continuous ponding? How many 1-ft high check dams would be required for a 200-ft long biofilter swale?

Setting Aside Vegetated Drainage Ways

As extensions of streams and drainage corridors, vegetated swales represent potential wildlife corridors, wetland multipliers of ecosystem integrity, recreational facilities close to home, and greenway links among neighborhoods.

The "blue-green" idea of urban development was articulated in the 1960s (Jones, 1967) to supply riparian open spaces where flood flows could spill over and be stored. Setting aside a vegetated drainage corridor accommodates a channel's natural tendency to aggrade, degrade, and meander. Channels flowing in vegetated corridors have shown less ongoing erosion than those encroached upon by structures (Whipple, DiLoie and Pytlar, 1981), perhaps because a channel that is allowed to find its own form is free to evolve a new equilibrium.

In the 1970s the floodway of Indian Bend Wash in Scottsdale and Tempe, Arizona, was set aside for recreational development and nature preserves, leaving capacity in the floodway to convey very large flows (Erie and Ueda, 1987). Destructive flooding had occurred repeatedly, but residents opposed a proposed seven-mile-long concrete channel that would achieve a 100-year capacity of 30,000 cfs with a single-purpose structure. The alternative plan achieved the same capacity with land use and landscape design that included a small low-flow channel in some constricted areas. Governmental and private agencies developed active and passive parks, wildlife sanctuaries, and trails. Everything that went into the 600- to 1,100-ft-wide floodway worked under hydraulic parameters so as not to obstruct the floodway without compensatory deepening or widening. The floodway now contains golf courses (with greens above flood level), sports fields and courts, fishing lakes, play and seating areas, and a trail system, all of which are heavily used in the months and years between floods.

Welsch (1991) distinguished the riparian ecological zones listed in Table 6.5. In riparian buffers, self-regenerating vegetation slows stream bank erosion and creates habitats for functioning aquatic and riparian ecosystems. It regulates shade, temperature, and nutrient balance and produces woody debris that structures aquatic habitats. Water flowing from the developed uplands is given time and room for attenuation and treatment in riparian soil and vegetation. The outermost zone in the table, the "runoff control" zone from which runoff is delivered to the riparian corridor, is the subject of the remaining chapters of this book. Runoffs must be delivered from urban sites with rate, volume, and quality such that the regenerative processes of the riparian corridor can absorb them and build them into the riparian ecosystem.

Natural Stream Models

Geomorphic restoration attempts to emulate the form and behavior of undisturbed streams. A natural stream is an erosional, transportational, and depositional system where form and process evolve together toward dynamic equilibrium.

TABLE 6.5 Functional zones around stream channels (Welsch, 1991, p. 11)

Distance from Stream Bank	Zone	Function and Required Management
0 to <15 feet	Undisturbed forest	Maintenance of stream habitat with shading and production of organic matter and woody debris. No disturbance permitted.
15 to <75 feet	Managed forest	Sediment and nutrient uptake and transformation. This zone must be wide enough to filter sediment from surface runoff; effective removal depends on uniform, shallow flow. Compatible management need not reduce the effectiveness of this zone.
≥75 feet	Runoff control	Runoff entering the stream corridor from outside must be converted to slow, nonerosive flow with vegetated swales, infiltration basins, or additional buffer width.

In steep headwater streams and swales, boulders, logs, leaves, and twigs wash into steps and pools. The steps dissipate energy. The pools provide low-velocity habitat; sediments that settle in the pools maintain stocks of nutrients. When a dam fails as a result of rotting, undermining, or washout, other debris takes its place, although not necessarily at the same location; steps and pools collapse and regenerate in a constantly shifting mosaic.

Where gradient is less, gravel beds form wavelike riffles while the channel meanders through floodplain alluvium. On the outside of each bend, fast-moving water undercuts the bank into a steep slope; on the inside, slowly moving water deposits sand and gravel into gently sloping point bars. Shallow, steeply sloping riffles are habitats for insects and centers of stream aeration. Deep, quiet pools are scoured out at the outside of bends and around deflecting logs and boulders; others are impounded upstream of obstructions; all are refuges for fish during low flows (Beschta and Platts, 1986).

Geomorphic restoration uses natural models that occur nearby in the region (Rosgen, 1996). Geomorphic principles can be used to predict the width, depth, sinuosity, and riffle-pool spacing that a channel would have if it were allowed to evolve into equilibrium with its given discharge and slope. Unlike uniformly cross-sectioned swales, geomorphically restored channels are meandering, interspersed with pools, and densely vegetated. Sequences of pools and riffles with abundant vegetation produce diverse habitats. Gentle grading at the sides allows point bars to develop.

Any proposal for stream alteration or management should be investigated for its potential flow and stability effects on upstream, downstream, and laterally adjacent areas. Each section of channel is unique. Sinuosity, slope, and width are interrelated; changing one causes the others to adjust. A measure to halt erosion at one place may induce it at another. Flexibility to adapt to stream features and behaviors as they evolve must be included in stream restoration planning.

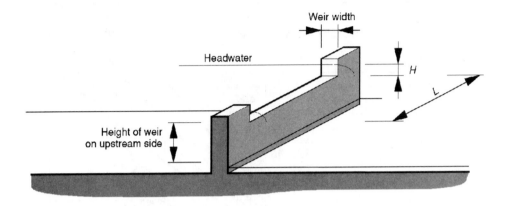

Figure 6.16   Dimensions of a rectangular weir.

WEIRS

A weir is a shelf that water falls over, such as the top of a barrel pipe, a timber check dam, or the crest of a concrete dam.

Weir discharge is governed by the simple principle that as head increases upstream of a weir, the flow over it increases. The type of material is irrelevant as long as the weir has only enough width for structural strength and water does not contact it long enough to be slowed down by friction.

A weir that is straight and level, with vertical sides, like that in Figure 6.16, is referred to as a rectangular weir. Discharge over this type of weir is predicted by the weir equation (U.S. Bureau of Reclamation, 1974, p. 373):

$$q = C L H^{3/2}$$

where

q = discharge, cfs

C = coefficient (no units) equal to 3.3 where a weir has significant height on the upstream side, such as at a dam or a barrel pipe, or 3.09 where a weir has no upstream height, such as at a drop structure in a channel

L = horizontal length of crest, ft

H = head of water above the crest, ft

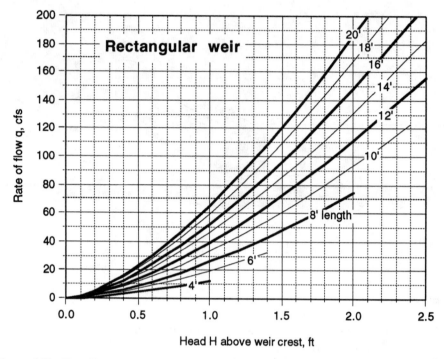

Figure 6.17 Flow in rectangular weir (based on weir equation with $C = 3.3$ and forward velocity over the weir = zero).

The preceding equation applies as long as length L is at least five times the head H; at shorter lengths or higher heads, the edges of the weir alter the rate of flow.

The chart in Figure 6.17 shows discharges through various lengths of rectangular weirs at C equal to 3.3. At a drop structure in a channel, the discharge would be 3.09/3.3, or 0.94, times the discharge shown on the chart. Rectangular weirs can be custom-built of concrete, masonry, or wood to fit any needed size; you can interpolate between the curves on the chart to derive a precise required length.

Flow over a circular weir, such as the top of a standpipe (Figure 6.18), is governed by the same weir equation. In this case the length L is equal to the circumference of the circle. The chart in Figure 6.19 shows discharges through commercially available diameters of pipe that may be used as barrel pipes. The chart is based on a full circle, with water approaching from all sides of the pipe. At a half-circle weir, such as a concrete weir installed as part of a dam, the discharge would equal half the discharge shown in the chart. The discharge over a weir with no height above the channel, such as a sill or drop structure, is 0.94 times the discharge given in the chart.

Grade Control

Weirs can be used to counteract the channel incision that commonly accompanies urbanization. A check dam (a small dam acting as a weir) raises the channel's base level and causes upstream aggradation. It reduces the upstream channel's gradient, velocity, and ero-

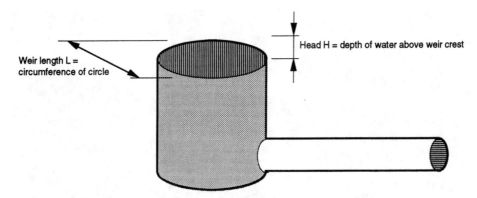

Figure 6.18   Circular weir formed by the top of a barrel pipe.

sive potential. In an incised stream, it reduces bank height and increases bank stability. An incised stream responds to check dams with storage of sediment and water, rising riparian groundwater, restoration of base flows, and stabilization of riparian habitats (Heede, 1977 and 1990). Application of grade control structures must be guided by analysis of the stream system and comparison with the channel's equilibrium form (Heede, 1986; Rosgen, 1996).

At Horseshoe Park in Aurora, Colorado, William Wenk's design established riparian wetlands by dropping the channel over a series of structures between which flood flows are

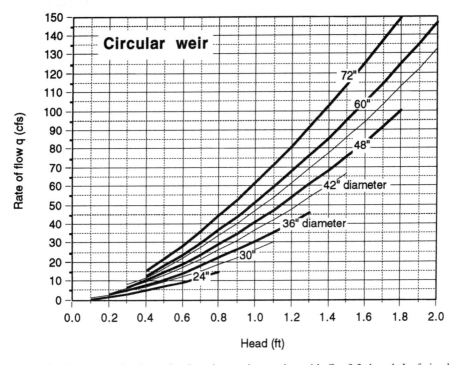

Figure 6.19   Flow over circular weirs (based on weir equation with $C = 3.3$; length L of circular weir = circumference of the circle).

Figure 6.20   Concrete check dam at Horseshoe Park, Aurora, Colorado.

allowed to spread out over the floodplain. Each structure is a couple of feet high and surrounded by rocks and miniature wetlands (Figure 6.20). Trails, pedestrian crossings, seating
areas, and numerous access points from the surrounding residential neighborhoods establish
the human use of the park land.

At Shop Creek, above the Cherry Creek Reservoir near Denver, Wenk molded a series of
soil-cement drop structures into crescents to blend with the rolling forms of the surrounding
prairie. Each structure establishes an upstream wetland to add wildlife habitat and remove
urban phosphorus from stream water.

At Strawberry Creek on the University of California campus in Berkeley (upstream
from the Strawberry Creek Park described in Chapter 2), check dams of rock and wood
were used to rehabilitate a long-distressed stream ecosystem (Charbonneau and Resh,
1992). The watershed had been grazed in the late nineteenth century and urbanized in the
twentieth. Erosion threatened trees and structures. Over the years the university had added
some concrete linings, dams, and retaining walls. Habitat loss, structural migration barriers,
and poor water quality devastated fish diversity. In the 1980s check dams were installed in
areas of steep channel gradient to prevent further downcutting. Dam height was limited to
allow fish migration; the low height necessitated close spacing. Most dams collected sediment wedges, stabilizing short reaches upstream. Downstream turbidity declined. Flow
over dams provided varied habitat for fish and insects. New and modified check dams
were accompanied by bank stabilization, sewer repair, and at-source pollution control.
Subsequently, native fish have been reintroduced, crayfish have immigrated from upstream,
large numbers of insects have colonized, and native egrets have been seen foraging for fish.
The principal impediment to further rehabilitation of the creek is the urban condition of the
watershed, which continues to cause flashy scouring flows and low base flow beyond the
mitigation capacity of in-stream treatments.

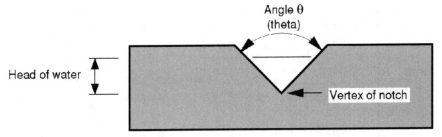

Figure 6.21   Dimensions of a vee-notch weir.

Vee-Notch Weir

A vee-notch weir (Figure 6.21) is useful for controlling very small rates of flow. At low head, the flow through the narrow notch near the vertex of the "vee" is quite slow. Rate of flow is predicted by (Debo and Reese, 1995, p. 134)

$$q = 2.5 \tan(\theta/2) \, H^{5/2}$$

where

q = rate of flow, cfs

θ = notch angle, degrees (θ is the Greek letter theta)

H = head above vertex, ft

The chart in Figure 6.22 shows rate of flow through various notch angles. Notch weirs can be custom-built of concrete, metal, or wood to fit any required angle. You can interpolate between the curves on the chart to derive a precise required angle.

ORIFICES

An orifice is an aperture through which water flows. Examples are the mouths of some culverts, holes specially cast in small concrete dams, and perforations cut into pipes.

Flow through a straight vertical orifice like that in Figure 6.23 is described by a simple equation,

$$q = C \, A \, (64.4 \, H)^{0.5}$$

where

q = rate of flow, cfs

C = coefficient equal to 0.8

A = area of the orifice, sf

H = head over the center of the orifice, ft

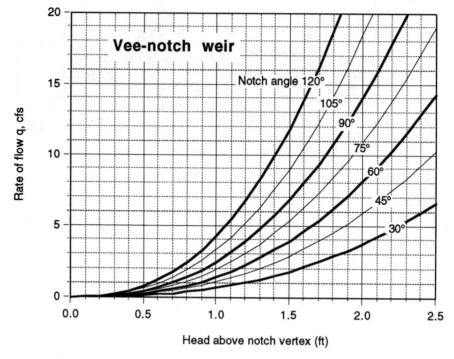

Figure 6.22 Flow in vee-notch weirs (based on vee-notch weir equation).

Figure 6.24 shows flow through circular orifices according to this equation.

A direct application of the orifice equation is an orifice plate. To control flow precisely, a hole is cut in a metal plate that is attached over the mouth of a culvert. The culvert carries away only the amount of flow that passes through the orifice.

A more complicated application of the orifice equation is a system of perforations cut or drilled in a standpipe. This arrangement is seen in the outlets of some detention and sedi-

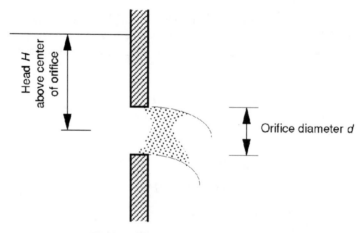

Figure 6.23 Dimensions of an orifice.

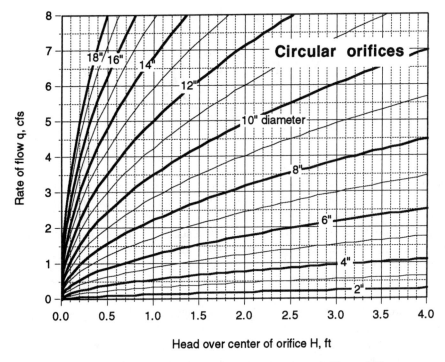

Figure 6.24 Flow in circular orifices like that shown in Figure 6.23.

ment-control ponds. Perforations of a given diameter are located in a grid pattern at a specified spacing. As water rises, it encounters higher rows of perforations, at the same time
increasing the head on lower rows. At any given water level, the discharge is the sum of the
flow through all the perforations, at their various heads.

PIPES

Pipes are used to convey stormwater under streets, under dams, and under the ground in
congested areas where open channels would not fit. Some cities convey all their stormwater along the streets in piped storm sewers. The adequate sizing of pipes is dear to the hearts
of municipal engineers, who receive the angry phone calls when a street is covered with
water from an overflowing storm sewer or culvert.

Pipes are usually more expensive than swales of similar capacity, because pipe materials
must be purchased whereas swales are primarily formed of earth and vegetated. But pipes
are essential where water has to be carried under roads or where site development is too
intense to leave room for broad earthen swales.

Figure 6.25 shows the cross section of a pipe. In hydraulic calculations references to the
diameter always refer to the inside diameter, because that is where the water flows.

The rate of flow q through a pipe is determined by the amount that passes through the
pipe's inlet. The mouth of a pipe is always a constriction in a drainage course; water
approaching it from an upstream swale does not necessarily enter the pipe at the same rate
it was flowing in the swale. Instead, a lot of the water backs up at the mouth and boils

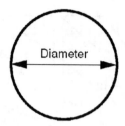

Figure 6.25 Diameter of a circular pipe.

around until it gathers itself into the pipe's entrance. The rate of flow into the pipe is then a function of depth of ponding (head) over the culvert's mouth and, thus, of pressure pushing the water in. This condition, called inlet control, is shown in Figure 6.26.

The flow through a pipe may be further constrained where there is significant depth of water at the outlet, backing up water into the pipe and slowing the throughflow. This could happen where a pipe discharges into a swale that is flowing full of water, or into a pond with a certain water level. In this condition the rate of flow into the pipe is determined by the elevations of water at the inlet and outlet and the horizontal distance between; the combination forms the hydraulic gradient that pushes the water through the pipe. This condition is called outlet control.

Manning's equation cannot be used to predict the rate of flow through a culvert. Manning's equation applies where flow is uniform—but the inlet of a pipe is an obstruction. The rate of flow through the pipe is determined by how much gets past the obstruction; Manning's equation does not tell you that. An unfortunate number of stormwater textbooks and manuals tell you to use Manning's equation to size pipes; they are wrong. Their assumption is that the capacity of the pipe to convey water is reached when the pipe is flowing full. But inlet and outlet control constrain the rate of flow before the pipe cylinder is flowing full. The water profile through the culvert in Figure 6.26 is typical. The pipe is full at the mouth, where water is backed up and the inlet is accepting flow at its capacity. After entering the pipe, the water gathers itself into forward flow down the pipe cylinder. The water surface declines as the water accelerates. In the middle of the pipe, the pipe is flowing only partly full even though it was flowing full at the mouth. Full flow occurs under only one condition: where the pipe is under such high outlet control that water is backed up along the pipe's entire length. You do not even know whether such a condition has occurred until you have analyzed the pipe for outlet control. Use inlet and outlet control to determine rate of flow through a pipe, not Manning's equation.

Manning's equation can be used to find velocity of flow in a pipe after the rate of flow has been determined. In Figure 6.26, in the middle of the pipe the flow has accelerated into

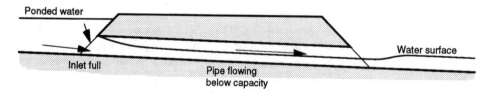

Figure 6.26 Water profile through a culvert with inlet control.

an equilibrium with the pipe's slope and roughness. Now that flow is unobstructed and uniform, Manning's equation applies to the given q, although it did not help you to find out what the value of q is.

Inlet Control

Inlet control applies where water is backed up at the mouth of a culvert but not at the downstream end. The head over the inlet makes it operate more or less like an orifice. Pressure and flow increase with increasing head. The entrance to every culvert should be checked for its inlet-control capacity.

The inlet control equation in a simplified form is (Normann et al., 1985, p. 146–147, assuming circular concrete pipe, submerged inlet, 2% slope, and square edge with headwall):

$$q = d^{2.5} \, (15.5 \, H / d - 10.2)^{0.5}$$

where

q = rate of flow, cfs

H = head above the invert at the pipe's inlet, ft

d = diameter of pipe, inches

You can see that the exponent of 0.5 is the same as that in the orifice equation. Other terms in the equation modify the orifice equation to measure head from the invert not from the center of the orifice. The equation is valid only where H is at least $1.3 \, d$.

Figure 6.27 shows the flow in inlet control through commercially available pipe sizes. To find the rate of flow through a given pipe, enter the chart from the bottom with the head at the inlet, move up to the line for the diameter of your pipe, thence to the left to read the rate of flow. To find the size of pipe required to convey a given flow, enter the chart from the bottom with the allowable depth of ponding at the inlet, and from the side with the rate of flow. If your lines intersect exactly on the curve for one of the pipe sizes, then that is your design size. If your lines intersect at a point between two of the pipe size curves, move up and to the left to find the next larger commercially available size.

Outlet Control

Outlet control applies where water is deep enough at the downstream end of a culvert to influence the rate of throughflow. This can occur where water is backed up by ponding or sluggish flow. It may not occur where the water is allowed to flow freely away from the outlet. In areas of level topography, such as parts of Minnesota and Florida, it is common practice to check the outlet of every culvert for its outlet-control capacity. Even in rolling topography, outlet control can occur where pipes and swales are laid out with low gradients.

To move through an outlet-controlled culvert, water has to push out at the outlet, as well as in at the inlet. The difference in elevation between headwater and tailwater and the horizontal length of the pipe establish the energy gradient that pushes water through the pipe.

Figure 6.28 shows the dimensions that produce the difference in head ΔH from headwater to tailwater. The difference can be found by using

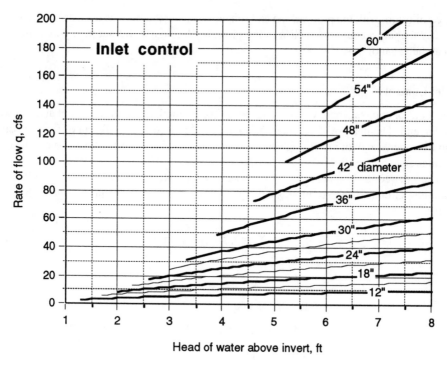

Figure 6.27 Flow in inlet control (from inlet control equation for circular concrete culvert at 2 percent slope, submerged inlet, square edge with headwall, on pages 146 and 147 of Normann et al., 1985).

$$\Delta H = D_e + D_i - D_o$$

where

ΔH = the difference in head between the headwater and tailwater surfaces, ft

D_e = drop in elevation of culvert invert from inlet to outlet, ft

D_i = depth of water at inlet, ft

D_o = depth of water at outlet, ft

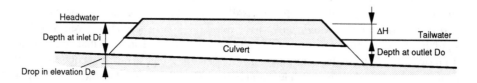

Figure 6.28 Outlet control dimensions.

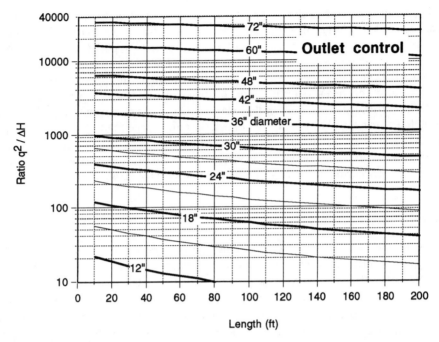

Figure 6.29 Flow in outlet control. Note that the vertical scale is logarithmic. The advantage of a logarithmic scale is that it makes small values as visible as large ones; however, it requires care in reading. (From outlet control equation for culvert submerged at both inlet and outlet and flowing full with square entry ($k_e = 0.5$ and $n = 0.013$), equation 5 on page 35 of Normann et al., 1985.)

Normann and others (1985, p. 35) developed a complicated equation for the rate of flow produced by this difference in head and the length of pipe. You can see the equation in their original publication.

The chart in Figure 6.29 implements this equation in the simplest way possible. It shows the ratio of q^2 and ΔH as a function of culvert length and diameter. To use the chart to find an adequate culvert size, first derive the head difference ΔH from the preceding equation, and obtain the rate of flow q required to be conveyed from a runoff estimate; calculate the ratio $q^2/\Delta H$ in cfs^2/ft. Enter the chart from the bottom with the length of the culvert, and from the side with the $q^2/\Delta H$ ratio. If your lines intersect exactly on the curve for one of the pipe sizes, then that is your design size. If your lines intersect at a point between two of the pipe size curves, move up and to the right to find the next larger commercially available size.

Velocity in Pipes

You need to know the velocity in a pipe that you are designing. Although concrete and metal pipes themselves cannot normally be abraded, the velocity discharging from a pipe can erode the soil in the receiving swale.

When q through a pipe has been determined from inlet or outlet control, it is possible to derive the velocity using Manning's equation. However, the mathematics for doing this are tedious, because in a partly full circular pipe the cross-sectional area and wetted perimeter, which are needed in Manning's equation, vary in a complicated way with q. So the equations are not presented here; they are left for brave students to derive in a term project. Instead, we go straight to charts that show the results of those equations.

Figures 6.30 through 6.34 show velocity as a function of q and slope. Concrete pipes are shown, because concrete's longevity is making this material increasingly popular as compared with corrugated metal. Each of the charts applies to one commercially available pipe size. Before using these charts, you have to find the pipe's slope from a topographic map or grading plan, and the q flowing through the pipe based on inlet or outlet control. To read each chart, enter from the bottom at the pipe's slope and move vertically up to the line for the correct q. Thence move horizontally to the left to read the resulting velocity.

Analytical Exercises

1. Is it possible for water to flow through a pipe when the pipe's slope is zero? What physical forces could cause this to happen? What kind of flow condition (Manning, inlet control, etc.) could you use to analyze the flow rate?

2. Is it possible for water to flow through a pipe when the end of the pipe is submerged below the permanent water level in a pond? What kind of flow condition could you use to analyze the flow rate? Are there any aesthetic or safety considerations that could cause you to design a pipe that way?

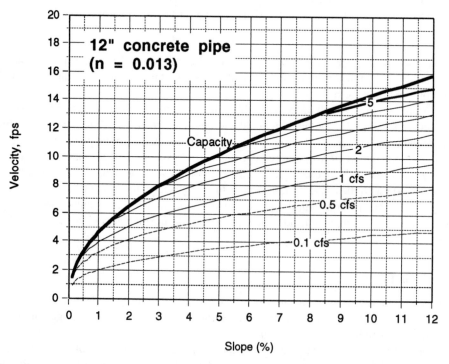

Figure 6.30 Velocity in 12-in. concrete pipe (based on Manning's equation and the geometry of a circular pipe).

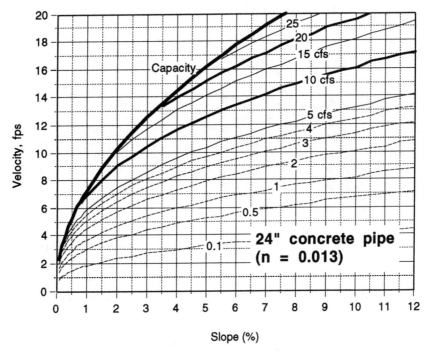

Figure 6.31 Velocity in 24-in. concrete pipe (based on Manning's equation and the geometry of a circular pipe).

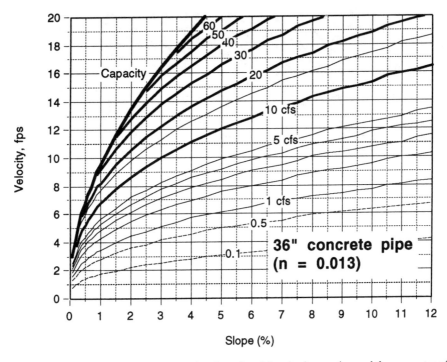

Figure 6.32 Velocity in 36-in. concrete pipe (based on Manning's equation and the geometry of a circular pipe).

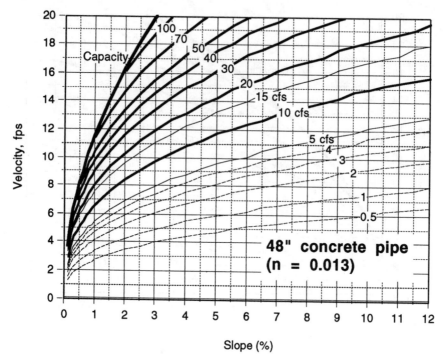

Figure 6.33 Velocity in 48-in. concrete pipe (based on Manning's equation and the geometry of a circular pipe).

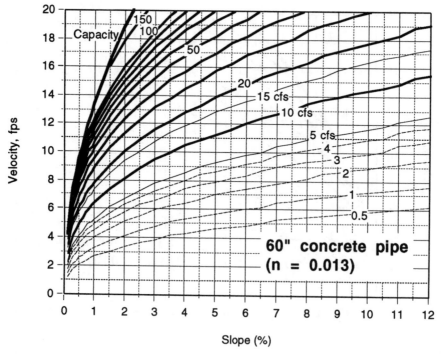

Figure 6.34 Velocity in 60-in. concrete pipe (based on Manning's equation and the geometry of a circular pipe).

3. Using algebra, derive from the geometry of a circle and $q = VA$ an equation for velocity of flow V through a full-flowing circular pipe, in terms of rate of flow q and diameter of pipe d.

4. This exercise is for an extended project. Derive the necessary equations describing the geometry of a circular pipe flowing partly full, and write a spreadsheet to derive the kind of data shown in the velocity graphs in Figures 6.30 through 6.34. *Hint*: A spreadsheet can include a look-up table.

Pipe Exercise

Exercise 6.2 contrasts pipe sizes found with inlet control and outlet control. The pipe discharges into the swale that was designed in Exercise 6.1. The exercise uses only the results of the SCS method. For the rest of this book only the SCS results will be used in exercises, because of the usefulness of the method's results for reservoirs and other applications where flow volume must be known.

Summary of Process

1. Obtain q_p after development from the SCS runoff estimate in Exercise 4.4.
2. Obtain the maximum noneroding velocity in the receiving swale from Exercise 6.1.

Design for Inlet Control

3. Specify maximum allowable headwater depth H at the inlet to the culvert, in feet.
4. Find the required pipe size from Figure 6.27 or the inlet control equation.

Design for Outlet Control

5. Specify length of culvert in feet.
6. Find slope of culvert in ft/ft from the site grading plan.
7. Find loss of invert elevation in pipe D_e from the site grading plan or by calculating length $\times$ slope.
8. Specify allowable depth at mouth D_i. This is the same as the allowable H for inlet control.
9. Find depth of flow at outlet D_o from the depth of flow in the swale you designed in Exercise 6.1.
10. Find head loss ΔH in feet using $\Delta H = D_e + D_i - D_o$.
11. Calculate the ratio $q_p^2 / \Delta H$, in cfs^2/ft.
12. Find required culvert diameter in Figure 6.29.

Pipe Selection and Check for Velocity

13. Select a pipe size adequate for both inlet and outlet control. This is the larger of the sizes found for the two possible control conditions.
14. Compute velocity through the pipe using a derivation of Manning's equation or find it in the charts in Figures 6.30 through 6.34.
15. If velocity exceeds the maximum allowable velocity, select a larger pipe, armor the erodible soil with a stable material, or revise the site plan to reduce the pipe's gradient or the rate of runoff reaching the pipe. Then begin again with step 1.

Exercise 6.2   Culvert. The culvert discharges into the swale designed with the SCS method in Exercise 6.1.

	Site 1	Site 2

Design data

q_p entering culvert, SCS method
   after development
   (from Exercise 4.4)   = _____ cfs   _____ cfs

Maximum noneroding velocity
   in receiving swale
   (from Exercise 6.1)   = _____ fps   _____ fps

Design for inlet control

Maximum allowable head at
   culvert's mouth (from site map) = _____ ft   _____ ft
Pipe size based on inlet control
   (from Figure 6.27)   = _____ in.   _____ in.

Design for outlet control

Length of culvert
   (from site map)   = _____ ft   _____ ft
Slope of culvert
   (from site map)   = _____ ft/ft   _____ ft/ft
Loss of invert elevation in pipe D_e
   = length × slope   = _____ ft   _____ ft
Allowable depth at mouth D_i
   (from site map)   = _____ ft   _____ ft
Depth of flow at outlet D_o
   (from Exercise 6.1)   = _____ ft   _____ ft
Head loss ΔH
   = $D_e + D_i - D_o$ = _____ ft   _____ ft
Ratio $q^2 / \Delta H$
   = $q_p^2 / \Delta H$ = _____ cfs^2/ft   _____ cfs^2/ft
Culvert diameter
   (from Figure 6.29)   = _____ in.   _____ in.

Pipe selection and check for velocity

Pipe size adequate for both inlet
   and outlet control
   (larger of the two)   = _____ in.   _____ in.
Velocity through selected pipe
   (from Figures 6.30–6.34)   = _____ fps   _____ fps

Discussion of Results

1. Which sizing method requires the larger culvert? Why? Larger pipes cost more money to build. Which size is the correct size? Why did the other method fail to yield the correct pipe size? Which sizing method would you use in practice?
2. Using an equation for q in a full-flowing pipe based on Manning's equation like that asked for in previous "Analytical Exercises," find a pipe size for your site based on Manning's equation. Is this size different from that found using inlet or outlet control? If so, which size would be more prudent to specify in your design, and why? Base this choice on the sizes alone, without considering the hydraulic theories behind them.

Systems of Pipes

Pipes are frequently connected into systems to drain large areas. As usual, every pipe in the system must be checked for both inlet and outlet control. But you do not know the depth at the outlet of each pipe until you have designed the next one downstream. Here is a procedure for moving through the whole system in an orderly manner, applying criteria commonly used in practice, to size every pipe in the system:

1. Lay out the system of pipes.
2. Find q_p in each pipe by routing storm flow through the system. For a sizable system, you will certainly use hydrologic software to do this.
3. Select a criterion for the allowable depth of ponding at the inlet and outlet of a culvert. For example, if 2 ft of cover is required over every culvert, then the allowable depth of ponding might be 2 ft greater than the diameter of a culvert, to confine ponding below the level of the adjacent earth cover.
4. Size each pipe for inlet control, beginning with the pipe farthest upstream and progressing downstream, applying the criterion for headwater depth that you selected in the previous step. You can do this using Figure 6.27 by entering the chart from the left with the known q and moving horizontally to the right, looking at the line for each pipe size until you find one where head does not exceed your criterion.
5. Check each pipe for outlet control, beginning at the lowest pipe and progressing upstream. For each pipe, assume that the maximum permissible tailwater depth is based on your ponding depth criterion for the next pipe downstream; this is consistent with the criterion you used for inlet control. Increase in size may be required for outlet control, as compared with that for inlet control, for some pipes.

Compound Conveyances

Compound conveyances are often needed to discharge a combination of different design storms at controlled rates, or to direct low flows to treatment systems and high flows separately to downstream discharge.

Figure 6.35 shows a weir overlying the inlet to a culvert. At low flows, discharge is controlled by the culvert. As the water level rises on the upstream side of the structure, the head at the weir would be zero until the water's surface reaches the weir's crest. Then a positive head would begin to operate at the weir, while the total head would still apply to the culvert.

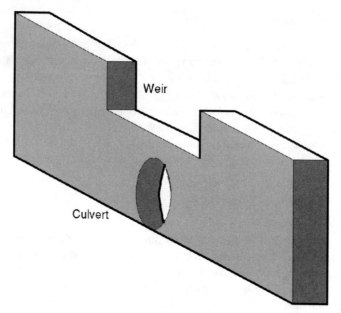

Figure 6.35 Culvert and weir making a compound conveyance.

The total flow through the compound conveyance is the sum of the flows through the culvert and the weir.

A complex example of a compound conveyance is a barrel pipe that is perforated to discharge low flows slowly, while large flows overtop the pipe altogether (Figure 6.36). Structures of this type have been used as outlets for detention and sedimentation basins. The barrel pipe is often discharged under a dam by a culvert attached to the bottom of the barrel pipe. At the junction of the two pipes an inlet or outlet control condition exists, because the water now has no forward velocity and its ability to enter the mouth of the culvert will determine how fast it can get going again. The culvert should be sized for either inlet or outlet control, whichever is more limiting.

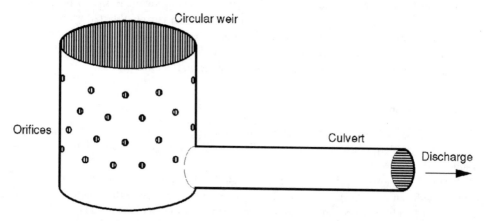

Figure 6.36 Perforated standpipe making a compound conveyance: orifices at lower elevations, weir at the top.

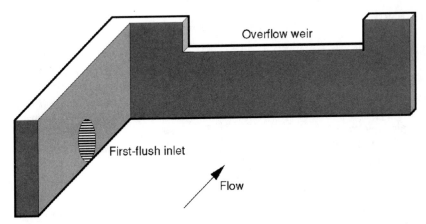

Figure 6.37   Compound conveyance used to split low first-flush flows for treatment away from high overflows.

A particular application of a compound conveyance is a "splitter" device. An example is shown in Figure 6.37. In the example, the inlet diverts low first-flush flows to a basin for treatment. High flows, in excess of the basin's capacity, discharge over the weir. Because the high flows do not pass through the basin, the basin is "off-line." The low flows being treated in the basin are not flushed out by high flows that exceed the capacity of the basin or that do not need treatment. Many configurations of splitters are possible, combining dikes to divert low flows and culverts and orifices to accept low flows at controlled rates.

SUMMARY AND COMMENTARY

Conveyance is a tool for managing flow. In some projects, it is the entire design objective. It is a hydraulic function for a site or a part of a site.

Conveyance addresses on-site nuisance abatement and safety. Appropriate site design for conveyance can address riparian ecology. Conveyance can also address, to a degree, concerns such as water quality, flooding, and stream base flows, whereby you specifically design swales and floodways for these purposes.

Preexisting channels and culverts are often included as parts of the conveyance systems in new developments. You can evaluate preexisting conveyances by using an equation or chart in reverse. In the appropriate chart, enter the known characteristics of a conveyance and work backwards to find the flow it can carry. Compare the result with the estimated peak flow of runoff that will enter the conveyance to see whether any reconstruction is necessary.

A variety of conditions may limit the rate at which water flows through a conveyance. Each conveyance must be designed for the runoff from its specific drainage area, and for its specific configuration, slope, and material.

But hydraulics alone need not dictate urban or environmental design. The results of conveyance calculations are a pipe's or channel's size, material, gradient, and velocity. These data are parameters for site design. They are not the answer to the site design problem; they are only constraints upon an answer. As long as you meet these few criteria, the hydraulic function will work, and in other respects you are free to design the site any way that is needed.

REFERENCES

Abt, Steven R., Rodney J. Wittler, James F. Ruff, and Mohammed S. Khattaak, 1988, Resistance to Flow over Riprap in Steep Channels, *Water Resources Bulletin* vol. 24, no. 6, pp. 1193–1200 (supplemented by personal communication from Steven R. Abt, September 12, 1989).

Arcement, George J., Jr., and Verne R. Schneider, 1989, *Guide for Selecting Manning's Roughness Coefficients for Natural Channels and Flood Plains*, Water-Supply Paper 2339, Washington: U.S. Geological Survey.

Barnes, Harry H., Jr., 1967, *Roughness Characteristics of Natural Channels*, Water-Supply Paper 1849, Washington: U.S. Geological Survey.

Beschta, R. L., and W. S. Platts, 1986, Morphological Features of Small Streams: Significance and Function, *Water Resources Bulletin* vol. 22, no. 3, pp. 369–379.

Charbonneau, Robert, and Vincent H. Resh, 1992, Strawberry Creek on the University of California, Berkeley, Campus: A Case History of Urban Stream Restoration, *Aquatic Conservation: Marine and Freshwater Ecosystems* vol. 2, pp. 293–307.

Chow, Ven Te, 1959, *Open-Channel Hydraulics*, New York: McGraw-Hill.

Corbett, Michael N., 1981, *A Better Place to Live: New Designs for Tomorrow's Communities*, Emmaus: Rodale Press.

Debo, Thomas N., and Andrew J. Reese, 1995, *Municipal Storm Water Management*, Boca Raton: Lewis.

Dunne, Thomas, and Luna B. Leopold, 1978, *Water in Environmental Planning*, San Francisco: Freeman.

Erie, Len, and James Y. Ueda, 1987, Indian Bend Wash: The Integration of Recreation, Flood Control, and Land Use, in *Computational Hydrology 87, Proceedings of the First International Conference*, Theodore V. Hromadka II and Richard H. McCuen, editors, pp. I-10–I-14, Mission Viejo, California: Lighthouse Publications.

Ferguson, Bruce K., 1991, Urban Stream Reclamation, *Journal of Soil and Water Conservation* vol. 46, no. 5, pp. 324–328.

Heede, Burchard H., 1977, *Case Study of a Watershed Rehabilitation Project, Alkali Creek, Colorado*, Research Paper RM-189, Fort Collins, Co.: U.S. Forest Service.

Heede, Burchard H., 1986, Designing for Dynamic Equilibrium in Streams, *Water Resources Bulletin* vol. 22, no. 3, pp. 351–357.

Heede, Burchard H., 1990, Conversion of Gullies to Vegetation-Lined Waterways: 26 Years Later, in *Erosion Control: Technology in Transition, Proceedings of Conference XXI*, pp. 99-113, Steamboat Springs, Co.: International Erosion Control Association.

Jones, D. Earl, 1967, Urban Hydrology—A Redirection, *Civil Engineering*, August, pp. 58–62.

Maddock, Thomas, Jr., 1976, A Primer on Floodplain Dynamics, *Journal of Soil and Water Conservation* vol. 31, no. 2, pp. 44–47.

Moulton, Lyle K., 1991, Aggregate for Drainage, Filtration, and Erosion Control, Chapter 12, *The Aggregate Handbook*, Richard D. Barksdale, editor, Washington: National Stone Association.

Municipality of Metropolitan Seattle, 1992, *Biofiltration Swale Performance, Recommendations, and Design Considerations*, Seattle: Municipality of Metropolitan Seattle Water Pollution Control Department.

National Stone Association, 1978, *Quarried Stone for Erosion and Sediment Control*, Washington: National Stone Association.

Normann, Jerome M., Robert J. Houghtalen, and William J. Johnston, 1985, *Hydraulic Design of Highway Culverts*, Hydraulic Design Series No. 5, Washington: U.S. Federal Highway Administration.

Rosgen, Dave, 1996, *Applied River Morphology*, Pagosa Springs, Co.: Wildland Hydrology.

Sykes, Robert D., 1988a, Channels and Ponds, *Handbook of Landscape Architectural Construction*, Volume 2, Maurice Nelischer, editor, pp. 239–317, Washington: Landscape Architecture Foundation.

Sykes, Robert D., 1988b, Surface Water Drainage, *Handbook of Landscape Architectural Construction*, Volume 1, Maurice Nelischer, editor, pp. 133–221, Washington: Landscape Architecture Foundation.

Thayer, Robert L., Jr., and Tricia Westbrook, 1990, Open Drainage Systems for Residential Communities: Case Studies from California's Central Valley, in *Proceedings of the 1989 Conference of the Council of Educators in Landscape Architecture*, Sara Katherine Williams and Robert R. Grist, editors, pp. 152-160, Washington: Landscape Architecture Foundation.

U.S. Agricultural Research Service, 1987, *Stability Design of Grass-Lined Open Channels*, Agriculture Handbook 667, Washington: U.S. Agricultural Research Service.

U.S. Bureau of Reclamation, 1974, *Design of Small Dams*, Washington: U.S. Bureau of Reclamation.

U.S. Federal Highway Administration, 1973a, *Design Charts for Open-Channel Flow*, Hydraulic Design Series No. 3, Washington: U.S. Federal Highway Administration.

U.S. Federal Highway Administration, 1973b, *Design of Roadside Drainage Channels*, Hydraulic Design Series No. 4, Washington: U.S. Federal Highway Administration.

U.S. Federal Highway Administration, 1975a, *Design of Stable Channels with Flexible Linings*, Hydraulic Engineering Circular No. 15, Washington: U.S. Federal Highway Administration.

U.S. Federal Highway Administration, 1975b, *Hydraulic Design of Energy Dissipators for Culverts and Channels*, Hydraulic Engineering Circular No. 14, Washington: U.S. Federal Highway Administration.

U.S. Soil Conservation Service, 1986, *Urban Hydrology for Small Watersheds*, Technical Release 55, Washington: U.S. Soil Conservation Service.

Welsch, D., 1991, *Riparian Forest Buffers, Function and Design for Protection and Enhancement of Water Resources*, Technical Publication NA-PR-07-91, Radnor, Pa.: U.S. Forest Service.

Whipple, William, Jr., J. M. DiLoie, and T. Pytlar, Jr., 1981, Erosional Potential of Streams in Urbanizing Areas, *Water Resources Bulletin* vol. 17, no. 1, pp. 36–45.

CHAPTER 7

DETENTION

Detention modifies conveyance to slow down of the rate of flow of surface runoff. It delays the passage of water during storm events in order to reduce flood peaks.

Urban detention started in the 1960s when it became known that development tends to be followed by increasing q_p and aggravated flood damage. The homes ruined, the industries closed, the roads shut off, and the bridges collapsed have motivated local and state governments to require preventive measures.

The target of storm detention is peak rate of runoff. To implement such a plan, one or more points on a development site or on the stream system must be identified where the effect of detention is to be measured. This may be a point where a major swale or stream leaves the site, a point farther downstream where the interaction of on-site and off-site flows can be evaluated, or a point downstream where the limited capacity of a pipe or channel or the presence of a sensitive floodprone structure or property makes peak flow rate a specific concern.

A common goal for detention is that q_p after development should not exceed that at the same place before development. In some locales the goal is that q_p not exceed that coming from the same drainage area containing a theoretical standard undeveloped land use, such as a meadow. In either case the q_p before development, which sets the standard for the detention design, is estimated using a storm runoff model.

Detention is accomplished with a basin or reservoir that temporarily stores storm runoff. Adding a detention basin to a development does not change the fact that development has altered the site. A basin insulates outside areas from the impact that has been created. By controlling peak outflow from a developed area it makes rivers downstream "think" that the development never really happened, at least as far as the single measure q_p is concerned (Figure 7.1).

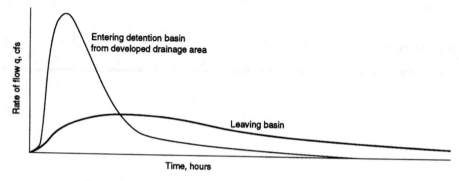

Figure 7.1   Effect of a detention basin on a storm hydrograph.

DETENTION BASINS

Figure 7.2 shows the elements of a detention basin. A constricted outlet retards the basin's outflow. Stormwater entering with high q_p backs up behind the outlet. The idea is to make the outlet small enough to reduce the flow to the required rate, and the basin big enough to store the difference.

The outlet is a conveyance. If it is a weir, the rate of flow through it is governed by the head over the crest. If it is a culvert, the rate is governed by inlet or outlet control. Protection from erosion is necessary at the discharge point.

A secondary overflow, or emergency spillway, is necessary to pass flows larger than that produced by the design storm without eroding the dam. Some spillways are earthen channels excavated at the side of the dam; others are large weirs or pipes higher than the principal outlet. Where the principal outlet is designed both to detain waters of the design storm and to pass very large flows, a separate emergency outlet may not be necessary.

The detention storage is the volume of the reservoir above the outlet's invert elevation. This is where runoff accumulates during storm events, to be let out gradually through the constricted outlet. This volume sits in reserve, full only of air, almost all the time. It is used for holding water only during flood events.

In a dry basin (Figure 7.3), the outlet's invert is flush with the basin's floor. When water backs up in the basin, the floor forms the bottom of the storage volume. Attempts have been made to get multiple use out of the floors of dry basins, but occasional ponding is a constraint that must be tolerated.

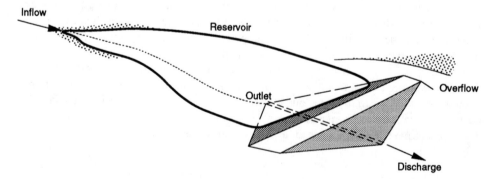

Figure 7.2   A detention basin.

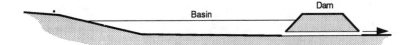

Figure 7.3   A dry basin; it empties completely following each storm.

A dry basin in Canton, Massachusetts, illustrates the kind of evolution to which most basins are susceptible. The basin was squeezed into a few hundred square feet in a development of town houses and apartments. It is bounded on one side by a stone-veneered retaining wall and on the others by steep slopes covered with cobbles. The basin is surrounded by grass and a canopy of native maple trees, and is overlooked by residential buildings. Two culverts drain into the basin; the outlet is a third culvert that passes under the road that bounds the property. There is often a small throughflow from the surrounding soil. Within a couple of years after construction, sediment began to fill in low points among the cobbles. Grasses and lowland plants are gradually growing up along the basin's narrow floor.

A more intensely developed and maintained dry basin is in Rio Rancho, New Mexico (Figure 7.4). Although it is structural and resource-consumptive, it is integrated with the life of its community. It is located in the midst of the houses, and is turfed and irrigated. At the upper elevations are trees, lighting, and play equipment. A concrete low-flow channel doubles as an actively used walkway. The dam's concrete-lined spillway is equipped and marked as a basketball play area.

Figure 7.5 shows a miniature detention area on a pavement. It is located at the back of a car repair shop, where the pavement is seldom used. The curb is high enough to contain a small volume of ponded water; the vee-notch in the curb controls the outflow. Water detained on a pavement has none of the benefits of contact with soil and vegetation.

Figure 7.4   A dry detention basin in Rio Rancho, New Mexico.

Figure 7.5   A detention area on a pavement in Athens, Georgia.

However, pavement detention offers double use of the space where commercial land is too valuable for a separate detention basin.

In a wet basin (Figure 7.6) a raised outlet establishes the water level of a permanent pool. When storm runoff enters the basin, the additional volume temporarily elevates the water surface. The volume of water permanently in the pool is not counted in detention storage because it is already preempted. A wet basin can be shaped and planted as a human amenity,

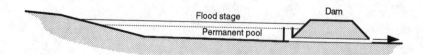

Figure 7.6   A wet basin; it includes a permanent pool.

Figure 7.7   Upper pond at the headquarters of General Telephone, near Columbus, Ohio.

as described in Chapter 2. A wet basin contributes to water quality in ways that a dry basin does not, as long as you design for it explicitly, using the criteria and procedures described in Chapter 8.

Near Columbus, Ohio, James Bassett designed two wet basins as amenities at the headquarters of General Telephone. The upper pond, shown in Figure 7.7, has little detention storage capacity. It is located directly outside the lobby, cafeteria, and other public rooms of the building and is ornamented with patios, paths, plantings and lighting. The second pond is the major stormwater storage pond; it has gravel edges to accommodate water-level fluctuations. It is located farther from the building and lower in elevation, and is surrounded by large grass areas and groves of trees, providing long vistas from the building across the large expanse of water. The ponds are connected by a rocky, carefully planted waterfall supplied by a pump that recirculates water from the lower pond to the upper.

In Davis, California, the West Davis Pond (Figure 7.8) is oriented more toward nature. This city-owned basin keeps discharges from storm sewers into a drainage canal within the canal's capacity. It was retrofitted as a wet basin to benefit wildlife. The earth was regraded into diverse islands and shallows and planted with native wetland vegetation. The outlet was raised to create the pool level in the wet season. In the dry season, the pool level is maintained with supplemental well water. Although the site is paralleled by a public greenway, a chain-link fence around the basin was motivated by safety concerns.

FLOOD STORAGE VOLUME

During a given design storm, to suppress peak flow to a given degree requires a certain definite amount of storage. Derivation of the required detention volume is based on the basic storage equation. During any increment of time,

Figure 7.8 West Davis Pond, a wet detention basin in Davis, California.

$$\Delta \text{storage} = \text{inflow} - \text{outflow}$$

In Figure 7.9 one curve shows the flow entering a detention basin; it represents inflow in the preceding equation. The other curve shows the flow of the same water leaving the basin; it represents outflow in the equation. In the early part of a storm, outflow is lower than inflow because of the constricted outlet. The difference between inflow and outflow is Δstorage; storage is accumulating as long as inflow is greater than outflow. It stops accumulating when inflow falls below outflow; the total accumulation up to that point is the maximum storage during the storm. The maximum storage is the required capacity of the basin; it is represented in the hydrograph by the area between the high inflow and low out-

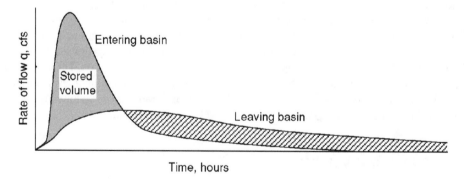

Figure 7.9 Detention storage volume determined by rates of flow entering and leaving the basin.

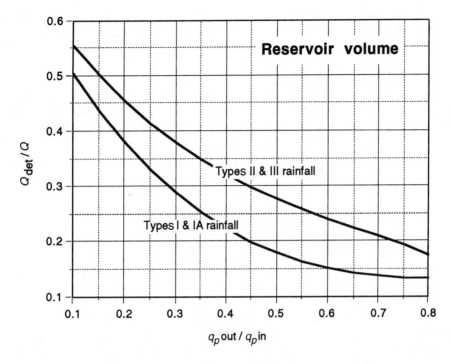

Figure 7.10 Detention storage volume as a function of suppression of peak flow rate (based on equation for Figure 6-1 given on page F-1 of U.S. Soil Conservation Service, 1986).

flow curves. After inflow falls below outflow, Δstorage is negative. The total volume discharging from storage is represented in the hydrograph by the area between the high outflow and low inflow curves; it equals the amount that had been stored earlier.

Reservoir routing is a thorough, precise way to estimate detention storage. With the use of this method, inflow, outflow, and Δstorage are computed for short increments of time and accumulated over the duration of the storm event. Although it can be done by hand, it is tediously time-consuming. Software that includes a reservoir routing module can do the calculations quickly, requiring from the user only accurate dimensions of the basin and its outlet and an inflow hydrograph to be routed through them.

Without a computer to do reservoir routing, detention volume can be estimated based on previous experience with calculations for other basins. An example is the chart in Figure 7.10. The scale on the chart's left side is the ratio of detention volume to runoff volume, Q_{det}/Q_{vol}. The scale at the bottom is the ratio of peak flow leaving the basin to peak flow entering, q_p out/q_p in. The curves in the chart show relationships between the two ratios found by SCS in numerous computer reservoir routings. There are separate curves for different SCS storm types, because timing of flow during a storm results in different storage regimes.

The chart can be easy to use as long as you spend a minute getting a feel for the intuitively reasonable principle on which it is based. The ratio $q_p$out/$q_p$in indicates how much "work" a basin is doing in suppressing peak flow. It is intuitively reasonable that if q_p is reduced greatly ($q_p$out/$q_p$in is low), then a large basin would be required to take up the large difference in flow. If q_p is reduced little ($q_p$out/$q_p$in is high), then only a small amount of storage is required. The slope of the curves across the chart from upper left to lower right expresses the simple relationship between the degree of suppression of peak flow and the amount of detention storage required to accomplish it.

To use the chart for design, first find the desired ratio of $q_p$out/$q_p$in from storm runoff estimates and local detention standards. Then enter the chart from the bottom at the desired ratio. Move up to the curve for the SCS rainfall type in your locale, thence horizontally to the left to read the required ratio Q_{det}/Q_{vol}. Obtain Q_{vol} entering the basin from a runoff estimate. The required detention storage capacity Q_{det} is found by using

$$Q_{det} = Q_{det}/Q_{vol} \times Q_{vol}$$

DETENTION LAYOUT AND OUTLET DESIGN

The exact layout of a reservoir to supply a given amount of storage depends on the opportunities of the specific site. On a cramped site you may be forced to make a basin deep and narrow to fit the available area. On a more generous site you might make it broad and shallow to blend with surrounding landforms and provide large areas of littoral vegetation. The required volume Q_{det} must be contained between the elevations of the outlet (at the bottom) and the maximum water level during the design storm (at the top), as illustrated in Figure 7.11. A grading plan with complete contours and spot elevations is necessary to document the volume that your basin provides. The volume that your grading plan produces can be conveniently evaluated with the contour planes method.

The design of the outlet depends on the stage (elevation) to which water rises, because outflow rate increases with head. The head that exists when detention storage is at its maximum is the difference in elevation between the outlet and the maximum stage of stored water.

From the conveyance charts in Chapter 6, select a type and size of outlet that produces the required outflow rate $q_p$out at the basin's maximum stage. Rectangular weir, vee-notch weir, and orifice outlets can be custom-built to any precise required size. However, if your outlet is a culvert or a standpipe, you are restricted to commercially available sizes; to match a given outlet size, you may have to recontour the basin to produce exactly the stage needed for the required outflow rate.

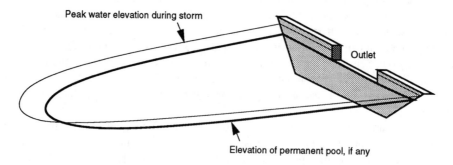

Peak water elevation during storm

Outlet

Elevation of permanent pool, if any

Figure 7.11 Changing water elevations in a detention basin.

Figure 7.12   The big basin at Coggins Park, Athens, Georgia.

THE STAGE-STORAGE-DISCHARGE RELATIONSHIP

Given the dimensions of a reservoir and its outlet, the flow of a flood wave through it can be analyzed in detail with the reservoir's stage-storage-discharge relationship. In any specific basin the stage, storage volume, and discharge (outflow rate) vary together.

An example of a stage-storage-discharge relationship is provided by the principal basin at Coggins Park, an industrial park in Athens, Georgia (Ferguson, 1993 and 1997; Ferguson and Gonnsen, 1993). Figure 7.12 shows the reservoir's outlet surrounded by ponded water following a storm. The outlet is a perforated standpipe wrapped in filter fabric to capture sediment. The top of the standpipe acts as a circular weir; it is surrounded by a trash rack.

Table 7.1 lists Coggins Park's stage-storage-discharge relationship. Stage is listed as elevation above sea level. During a storm, stage rises. It subsides to its original level afterward.

TABLE 7.1   Stage-storage-discharge relationship for the principal basin at Coggins Park, Athens, Georgia (Ferguson, 1993, p. 55)

Stage (ft)	Storage (af)	Discharge (cfs)	Comments
695.5	0.0	0.0	Basin floor
696.0	0.4	0.1	Perforations in operation
698.0	2.8	0.8	Perforations in operation
699.2	4.4	1.4	Perforations in operation; top of standpipe
700.0	5.6	39.1	Perforations + weir in operation
702.0	9.2	246.3	Perforations + weir in operation; spillway invert
703.0	11.4	454.2	Perforations + weir + spillway in operation

At each listed stage, storage was calculated by the contour planes method from a topographic map with 5-feet contours. Figure 7.13 shows the result. As stage increases during a storm, storage also increases. The basin is located in a steep narrow valley, so storage rises only slowly with stage; deep ponding is necessary to produce significant storage.

Discharge was calculated with the orifice and weir equations at appropriate stages. Figure 7.14 shows the results. As stage rises, head and discharge rise with it. At all stages, the pipe's perforations act as orifices with little discharge. Above 699.2 feet, the weir goes into operation too, adding its discharge to that of the orifices. Above 702.0 feet, the emergency spillway is in operation, so discharge is the sum of the flows through the orifices, over the weir, and through the spillway.

Because storage and discharge are both related to stage, they are related to each other. Figure 7.15 shows their connection at Coggins Park. During a storm the maximum discharge occurs simultaneously with the maximum storage. That maximum moment is the subject of detention design.

Reservoir routing makes full use of detailed stage-storage-discharge data to confirm estimates of required storage, such as those derived from Figure 7.10, and to produce a detailed discharge hydrograph that can be evaluated for its effects on erosion and its continued routing downstream.

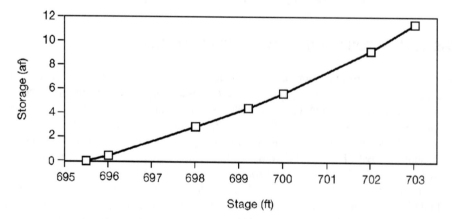

Figure 7.13   Stage-storage curve at Coggins Park.

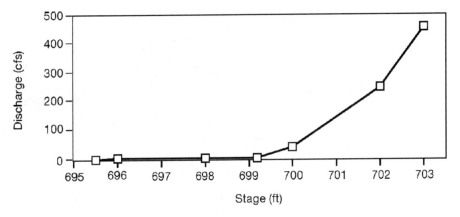

Figure 7.14   Stage-discharge curve at Coggins Park.

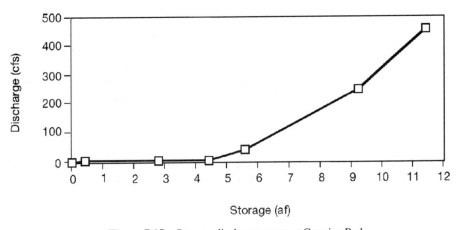

Figure 7.15   Storage-discharge curve at Coggins Park.

DETENTION FOR MULTIPLE DESIGN STORMS

In some locales two or more design storms are required for flood control. A basin designed for two design storms is diagrammed in Figure 7.16. Here are two approaches to meeting the requirements of two or more design storms:

1. Design the volume and outlet first for the largest storm. Develop a stage-storage-discharge table like that in Table 7.1. Then use the table with Figure 7.10 or a reservoir-routing program to determine the stage, volume, and discharge that will occur when the smaller storm comes through. Because the head at the outlet is lower during smaller storms, the outflow rate is lower. In some cases it happens that an outlet designed for the largest storm also controls smaller storms adequately. If not, possibly reducing the size of the outlet for the largest storm, thereby overcontrolling the largest storm, could create a basin that works adequately for smaller storms as well. Any reduction in outlet size is accompanied by greater stage and storage, because outflow rate is reduced.

2. Design a multistage outlet. This is a compound conveyance, the concept of which was introduced in Chapter 6. Use Figure 7.10 or a reservoir routing program to determine

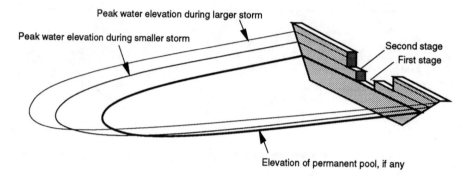

Peak water elevation during larger storm

Peak water elevation during smaller storm

Second stage
First stage

Elevation of permanent pool, if any

Figure 7.16 Changing water elevations in a multistage detention basin.

the volume required for the smallest storm. From the grading plan, find the stage to which water rises to produce this volume. Design an outlet for the smallest design storm using the appropriate conveyance chart or equation. At the stage to which water rises during the smallest storm, set the invert elevation of the outlet's second stage. Use the volume chart in Figure 7.10 or a reservoir routing program to find the storage, and thus stage, to which water will rise during the larger storm. During the larger storm, the head over the first (lower) stage of the outlet is given by

$$\text{Head} = \text{water elevation} - \text{outlet invert elevation}$$

The flow through the first stage during the larger storm can then be found from the appropriate conveyance chart. The remaining flow to be released by the upper stage of the outlet is

$$\text{Remaining } q = \text{total } q - q \text{ through first stage}$$

You can then design the second stage of the outlet to pass this rate of flow.

DETENTION EXERCISE

Exercise 7.1 designs a detention basin in which q_p after development must not exceed q_p before development and compares different outlet configurations. It uses the storm runoff estimates derived from the SCS method, because detention design requires an estimate of flow volume. Use the same sites that you used in Exercises 4.1 through 4.4.

Summary of Process

1. From the SCS runoff estimates in Exercises 4.3 and 4.4, obtain q_p before and after development, Q_{vol} after development, and SCS rainfall distribution type.

   Detention Volume

2. Compute the peak flow ratio (cfs/cfs):

$$q_p\text{out}/q_p\text{in} = q_p \text{ before development} / q_p \text{ after development}$$

Exercise 7.1   Storm detention.

	Site 1	Site 2

Determining data

q_p before development
(from Exercise 4.3) = _____ cfs   _____ cfs
q_p after development
(from Exercise 4.4) = _____ cfs   _____ cfs
Q_{vol} after development
(from Exercise 4.4) = _____ af   _____ af
Rainfall distribution type
(from Exercise 4.4) = _____   _____
Depth at outlet H
(from grading plan) = _____ ft   _____ ft

Detention volume

Peak flow ratio $q_p$out/ $q_p$in
= q_p before / q_p after = _____ cfs/cfs   _____ cfs/cfs
Volume ratio Q_{det}/ Q_{vol}
(from Figure 7.10) = _____ af/af   _____ af/af
Required storage volume Q_{det}
= $Q_{det}/ Q_{vol} \times Q_{vol}$ = _____ af   _____ af
Approximate area of basin
= Q_{det} / H = _____ ac   _____ ac

Outlet constraints

Head at outlet
= depth H = _____ ft   _____ ft
Rate of flow through outlet q_p out
= q_p before development = _____ cfs   _____ cfs

Rectangular weir outlet

Weir length L
(from Figure 6.17) = _____ ft   _____ ft

Vee-notch weir outlet

Notch angle θ
(from Figure 6.22) = _____ °   _____ °

Orifice outlet

Orifice diameter
(from Figure 6.24) = _____ in.   _____ in.

3. Read the volume ratio (af/af) from the chart in Figure 7.10.

4. Compute required storage volume Q_{det} in af:

$$Q_{det} = Q_{det} / Q_{vol} \times Q_{vol}$$

Detention outlet

5. Lay out the basin on the site grading plan in a way that satisfies the detention volume requirement. From the plan, read water depth (head) H above the outlet's invert elevation.
6. Set flow through the outlet $q_p$out equal to peak rate of flow before development.
7. For a rectangular weir outlet, read the required length L from Figure 6.17.
8. For a vee-notch weir outlet, read the required notch angle θ from Figure 6.22.
9. For an orifice outlet, read the required orifice diameter from Figure 6.24.
10. Check your results by routing the design storm through your basin using a computer program, if available.

Discussion of Results

1. Which site requires the larger detention volume? What factors contributed to the requirement of such a large basin? Does a great amount of natural (before-development) runoff necessarily indicate that a large detention basin will be required after development? Why?
2. What is the proportion of each site that must be occupied by a detention basin (basin area ÷ total site area)? In your judgment, does either basin occupy too much land to be reasonable purely in terms of land use?
3. If your response to the last part of question 2 was yes, some approaches you might take to solving the problem include changing the development's type of land use, choosing another site for this type of development, using underground storage chambers, reducing the area of impervious surfaces, and modifying the standard for detention control. What approach would you most likely choose? What specific site features would you have to build in order to implement your approach?
4. The peak outflow from a detention basin always coincides with a point on the receding limb of the inflow curve. Using algebra, derive that principle from the basic storage equation, Δstorage = inflow − outflow.

DETENTION'S WATERSHED-WIDE EFFECTS

At the exact spot where a detention basin discharges through its outlet, it reduces the peak rate of storm flow. We know this conclusively; the laws of physics compel it. But farther downstream, a basin's effect on peak rate depends partly on how its discharge combines with the flows from other tributaries. In practice, on any given site, detention should be applied only with caution.

Consider the watershed in Figure 7.17. A small development site discharges into the main stream of a large watershed. As shown in Figure 7.18, the storm hydrograph from the development site is short and fast as compared with that from the main watershed. Because the development site's flow drains out before the main watershed's peak arrives, it does not contribute to the magnitude of a flood downstream. But if detention is added to the developed site, outflow will be delayed, so that it overlaps onto the peak flow in the main stream and contributes to a new, higher combined peak flow.

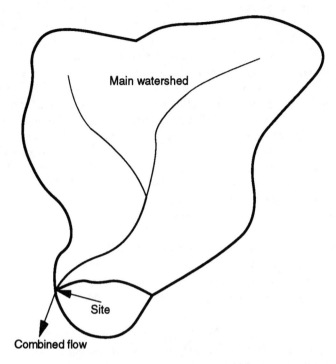

Figure 7.17 A watershed where the drainage from a small development site joins the flow from a large watershed.

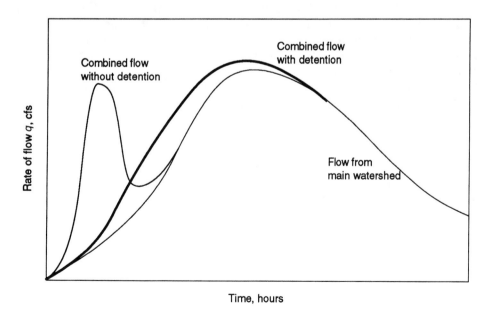

Figure 7.18 Alternative hydrographs from the watershed shown in Figure 7.17.

Now imagine two detention basins on different sites in the same watershed, constructed by different developers at about the same time. When the hydrographs from the two basins combine downstream, their delayed flows combine in a way that had never existed before development, and a larger flood may be created.

Nevertheless, numerous local governments have required every developer to reduce the peak rate during a design storm to its predevelopment level. The effect of this practice has been a random proliferation of small detention basins over urbanizing watersheds, none of which is designed with regard to its specific location in the drainage network.

The potential conflict between a basin and its watershed was pointed out as early as 1979, for a Maryland watershed (McCuen, 1979). Since that time, independent modeling studies in Colorado, Georgia, and Virginia (summarized by Ferguson, 1991, and supplemented subsequently by Ferguson, 1995, and Hess and Inman, 1994) have confirmed that randomly sited detention basins have failed. Enough studies have been done that the following generalizations can be drawn from them:

- Most watershed-wide systems of detention basins help, in the sense that they keep downstream q_p during a given storm lower than it would be without them.
- Some individual basins—perhaps 5 or 10 percent of them—do the opposite of help; they increase q_p downstream as a result of the overlapping of their detained volumes with mainstream peaks.
- No watershed-wide system of uniform basins works to the extent for which they were designed. If they were designed to reduce q_p during a given design storm to predevelopment levels, then their aggregate effect, although it is usually a reduction in q_p, is not a reduction to the designed degree, because of the accumulation of runoff volumes downstream.

A detention basin does not eliminate runoff; it delays it. The volume discharging from a basin is the same as the inflow. When the after-development volumes from different tributaries join downstream, there is nothing to prevent them from combining to produce inadvertently high peak rates. In the lucky cases in which flood peaks are consistently reduced, the receiving streams may still erode, because in accommodating the great volume of runoff, relatively high erosive flows still pass through for longer periods (McCuen, 1987). On a watershed-wide scale, uniform detention is a failure. It has no favorable effect on base flows, does not necessarily do anything for water quality, and fails to fulfill its single explicit purpose of controlling floods.

Detention basins can reduce flood peaks—when they are selectively located in their watersheds. Selective planning of publicly financed reservoirs led to effective flood control for the Miami River in Ohio in the early years of the twentieth century, when the Miami Conservancy District (Morgan, 1951) identified specific flood hazards in Dayton and other cities, then located a combination of multiple-purpose reservoirs, levees, and channels to work in concert to reduce flood damage at those points.

The potential effect of every proposed basin should be evaluated on a site-by-site basis. The practical way to check downstream effects is to use a computer program that routes flood hydrographs from different tributaries through the drainage system. Take your analysis as far downstream as the effects of your development may significantly extend, so you can see the results of your proposed plans in every segment of every stream.

Analytical Exercises

1. What constraints do you think should be placed on implementing detention on a regional or watershed scale? What criteria would you use for the location and design of detention basins? If you were an administrator of such a regional program, how would you make sure that the right thing is, in fact, being done on each site as it is developed?

    The remaining exercises require hydrologic software. The great speed of a computer at doing hydrologic calculations makes it possible to answer "what if" questions quickly and precisely, after basic data have been entered for a given site. The following questions assume that a detention basin been proposed in a watershed and that a routing calculation has been done using one storm recurrence interval.

2. What is the effect on discharge, basin area, and basin volume of moving the detention basin to a different location in the drainage system?

3. What is the effect on the storm hydrograph of adding more than one detention basin in series in the same watershed? What is the effect on the area and volume of detention basins, totaling all basins together?

4. What is the effect on discharge, basin area, and basin volume of increasing the impervious area in each subwatershed? You can assume that the pervious area is reduced by half. Then assume that the pervious area is eliminated entirely, leaving 100 percent impervious cover. What are some land use types that might involve 100 percent impervious cover?

5. What is the effect on discharge, basin area, and basin volume of using a significantly different recurrence interval for your design storm? If you have done calculations using a 10-year storm, try a 100-year storm and a 2-year storm.

SUMMARY AND COMMENTARY

Storm detention is a modification of conveyance; it is a modified scenario for moving surface water around and discharging it from a development site.

Detention reduces the peak rate of flow discharging from a site. It does not affect total flow volume. The entire volume eventually outlets from a detention basin and continues downstream. Storm detention does not, by itself, address water quality, groundwater, base flow, channel erosion, or aquatic and riparian habitat.

A detention basin is a facility in a site development program. It must be given adequate space in a site plan and integrated with its community.

The results of hydrologic calculations for storm detention give the required basin size and outlet configuration. As long as you stay within these few parameters, you are free to design the site any way that is needed for the environment and for quality of human life.

In some locales it is appropriate to remove the detention function from individual development sites. Developers sometimes pay an "impact fee" rather than build an on-site basin. The fee goes to a local agency, which uses its revenue to build regional flood control reservoirs at selected points in the river system. Regional reservoirs can be designed as public parks and maintained by the park agency. However, such a regionalized program sacrifices the unprotected headwaters of streams to increased urban runoff, and the total volume of runoff is still unreduced.

REFERENCES

Cavacas, Alan, 1985, Regional Watershed Management: Some Unanswered Questions, in *Proceedings, 1985 International Symposium on Urban Hydrology, Hydraulic Infrastructures and Water Quality Control*, pp. 109–120, Lexington: University of Kentucky.

Debo, Thomas N., 1982, Detention Ordinances: Solving or Causing Problems? in *Proceedings of the Conference on Stormwater Detention Facilities*, William DeGroot, editor, pp. 332–341, New York: American Society of Civil Engineers.

Debo, Thomas N., and George E. Small, 1989, Detention Storage: Its Design and Use, *Public Works*, January, pp. 71–72.

DeGroot, William, editor, 1982, *Proceedings of the Conference on Stormwater Detention Facilities*, New York: American Society of Civil Engineers.

Ferguson, Bruce K., 1991, The Failure of Detention and the Future of Stormwater Design, *Landscape Architecture* vol. 81, no. 12, pp. 76–79.

Ferguson, Bruce K., 1993, *Stormwater Management Ordinance Compliance, Coggins Park, February 1993*, Athens, Ga.: Beall, Gonnsen and Company.

Ferguson, Bruce K., 1995, Downstream Hydrographic Effects of Urban Stormwater Detention and Infiltration, in *Proceedings of the 1995 Georgia Water Resources Conference*, Kathryn J. Hatcher, editor, pp. 128–131, Athens: University of Georgia Institute of Government.

Ferguson, Bruce K., 1997, The Alluvial Progress of Piedmont Streams, in *Effects of Watershed Development and Management on Aquatic Ecosystems*, Larry A. Roesner, ed., pp. 132–143, New York: American Society of Civil Engineers.

Ferguson, Bruce K., and P. Rexford Gonnsen, 1993, Stream Rehabilitation in a Disturbed Industrial Watershed, in *Proceedings of 1993 Georgia Water Resources Conference*, Kathryn J. Hatcher, editor, pp. 146–149, Athens: University of Georgia Institute of Natural Resources.

Hess, Glen W., and Ernest J. Inman, 1994, *Effects of Urban Flood-Detention Reservoirs on Peak Discharges in Gwinnett County, Georgia*, Water-Resources Investigations Report 94-4004, Atlanta: U.S. Geological Survey.

McCuen, Richard H., 1979, Downstream Effects of Stormwater Management Basins, *Journal of the Hydraulics Division, Proceedings of the American Society of Civil Engineers* vol. 105, no. HY11, pp. 1343–1356.

McCuen, Richard H., 1987, Multicriterion Stormwater Management Methods, *Journal of Water Resources Planning and Management* vol. 114, no. 4, pp. 414–431.

Morgan, Arthur Ernest, 1951, *The Miami Conservancy District*, New York: McGraw-Hill.

Nix, Stephan J., and S. Rocky Durrans, 1996, Off-Line Stormwater Detention Systems, *Water Resources Bulletin* vol. 32, no. 6, pp. 1329–1340.

Poertner, H. G., 1974, *Practices in Detention of Urban Stormwater Runoff*, Special Report No. 43, Chicago: American Public Works Association.

Sykes, Robert D., 1988, Channels and Ponds, *Handbook of Landscape Architectural Construction*, Volume 2, Maurice Nelischer, editor, pp. 239–317, Washington: Landscape Architecture Foundation.

U.S. Soil Conservation Service, 1986, *Urban Hydrology for Small Watersheds*, Technical Release 55, second edition, Washington: U.S. Soil Conservation Service.

CHAPTER 8

EXTENDED DETENTION

Extended detention aims at improving water quality. In the still water of a pond, solid sediment particles and the pollutants attached to them settle out and microbiota may begin to degrade some dissolved constituents. Improving water quality requires a longer ponding time than flood-control detention, hence the name "extended" detention. Interest in this process grew markedly when EPA imposed a runoff quality mandate on local governments (U.S. Environmental Protection Agency, 1990). An additional spur to its use has been its potential contribution to wetland creation and enhancement. Because the approach is widely believed to require a permanent pool, it is known in many areas as wet detention; in some other areas it is known as stormwater wetland or constructed wetland.

The design storm for extended detention needs to be only a small, frequent storm. Water quality in the environment is a function of the pattern created by frequent events. For example, a uniform runoff volume of ½ inch is used in some locales; in others the two-year storm and the six-month storm are used. Designing to improve the water quality during a large storm such as a flood-control design storm would increase basin size and cost out of proportion to the benefit gained.

A general measure of water-quality control is trap efficiency. For a given type of constituent, trap efficiency is the ratio of the amount retained in a pond to the amount flowing in. It can be derived from measurements of inflow and outflow over a given period of time with the use of

$$E = 1 - L_o / L_i$$

where

$$E = \text{trap efficiency, kg/kg}$$

$$L_o = \text{load of constituent in outflow, kg}$$

$$L_i = \text{load of constituent in inflow, kg}$$

THE EXAMPLE OF LAKE ELLYN

A case study from the Chicago area illustrates how constituents are trapped in a pond. Lake Ellyn is a permanent pool that was created in 1889 by the construction of a dam on a tributary of the East Branch Du Page River. It is an urban lake: 80 percent of its watershed is single-family residential; the remainder is a combination of multifamily, commercial, park, and other urban land uses. Some of the lake's characteristics are listed in Table 8.1. From 1970 to 1980 the U.S. Geological Survey (Striegl, 1987) monitored the lake's inflow and outflow and the accumulating sediment. Inflowing sediment originated mostly from soil erosion, decomposition of paved surfaces, and traffic-related sources. Inflow commonly included pieces of glass, metal, and construction and packaging materials.

Figure 8.1 shows the mean particle size of the deposited sediment. Most of the lake floor was covered with an organic-rich mud. Silt and sand were concentrated in only in a few places near inlets. Grading away from the inlets, mean particle size became smaller, indicating that fine particles stayed suspended in the water longer. Although sediment near inlets had oil coating and odor, the highest concentrations of trace metals from motor vehicles and other sources were in the finest sediment deposits farthest

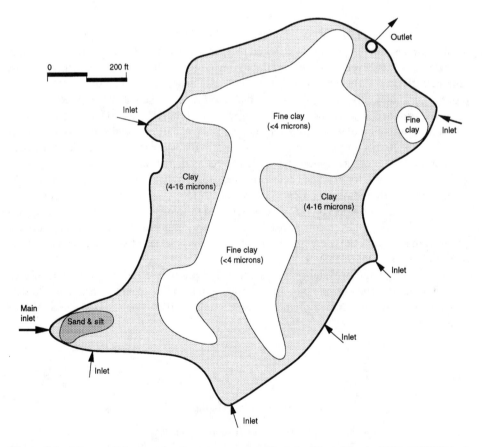

Figure 8.1 Mean particle size of sediment accumulated in Lake Ellyn, Illinois, 1970–1980 (Striegl, 1987).

TABLE 8.1   Characteristics of Lake Ellyn, Illinois (Striegl, 1987)

Impervious watershed coverage	34 percent
Drainage area at main inlet	73 percent of total drainage area
Lake area	10.1 acres
Maximum depth	6.5 feet
Permanent pool volume (equivalent)	1.0 inch of watershed runoff

from the inlets. The pond removed (trapped) 76 to 94 percent of the metals in the inflowing runoff.

Figure 8.2 shows the depth of sediment accumulation. The pond trapped 91 to 95 percent of the suspended sediment that entered it; the average accumulation on the bottom was 0.8 inches per year. Sand settled deeply in the main inlet's forebay. From each major inlet a plume of concentrated sediment deposition extended into the pond, getting shallower with distance from the inlets. The plume from the main inlet pinches out halfway down the length of the pond. The plume from the northeasternmost inlet made it all the way to the outlet, because the distance is not long, indicating that some sediment from this inlet was still in suspension when it discharged through the outlet.

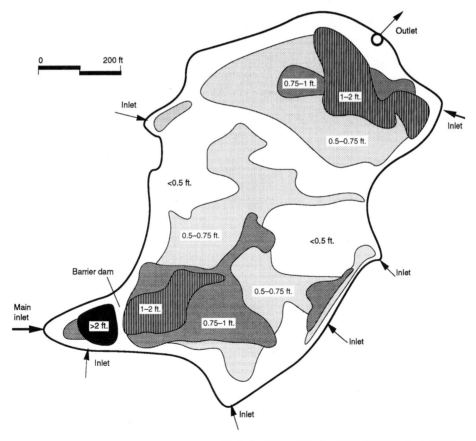

Figure 8.2   Depth of sediment accumulated in Lake Ellyn, Illinois, 1970–1980 (Striegl, 1987).

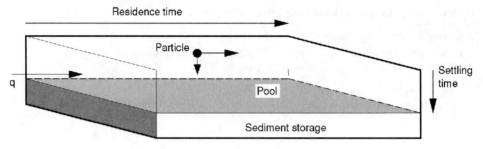

Figure 8.3 Physical settling.

PHYSICAL SETTLING

Figure 8.3 shows how a solid particle settles in this kind of environment. A suspended particle has a forward velocity resulting from the rate of flow q through the pool. The particle also has a vertical settling velocity. A particle effectively settles out of the water when it reaches the bottom of the pool. In order for the particle to do so before it is discharged from the basin, its residence time in the pool must be at least as long as the settling time.

Residence time is the time water spends in the pond's still-water environment while flowing from the inlet to the outlet. Average residence time is defined by

$$t_r = Q_p / q$$

where

$$t_r = \text{residence time, days}$$
$$Q_p = \text{volume of water in the pool, af}$$
$$q = \text{inflowing runoff, af/day}$$

Settling time t_s in days is given by

$$t_s = D_p / V_s$$

where

$$D_p = \text{pool depth, ft}$$
$$V_s = \text{settling velocity, ft/day}$$

Settling velocities of mineral particles are listed in Table 8.2. Gravel and coarse sand settle with visible speed because of their concentrated weight. Fine sand and silt take longer. Clay can take most of a year to settle one foot. Colloids—particles smaller than ordinary clays—take most of your lifetime to settle one foot. A value of settling velocity that has sometimes been selected for use in design is 1.0 ft/day; this is the velocity of a particle intermediate in size between silt and clay.

Effective settling is easier to accomplish with sand and gravel than with clay, because of the coarse particles' rapid settling velocities. However, large portions of chemical con-

TABLE 8.2   Settling velocities of mineral particles in still water (Van der Leeden, Troise, and Todd, 1990, p. 88)

Particle Size	Settling Velocity (ft/day)	Time to Settle 1 ft
Gravel	283,000	0.3 seconds
Coarse sand	28,300	3.0 seconds
Fine sand	2,260	38.0 seconds
Silt	43.6	33.0 minutes
Clay	0.00436	230 days
Colloids	0.0000436	63 years

stituents are carried by adsorption on clay particles; the chemical constituents are relatively hard to capture by physical settling.

DRY BASINS

A dry extended-detention basin empties completely after every storm and remains empty until the next storm starts. An undersized outlet produces a temporary pool, where settling takes place. The size of the outlet determines the time the pool is in existence while drawing down at the end of the storm.

Figure 8.4 shows a dry basin near the Baltimore airport. The ponding area is behind the concrete dam. A small PVC pipe cast into the base of the dam is the extended-detention outlet. The weir at the top of the dam is the overflow.

Figure 8.4   Dry extended-detention basin near the Baltimore airport.

Dry basins are effective to a degree. Wigginton and others (1983) found trace metals accumulated in the surface soils on the floors of dry basins near Washington, D.C. The metals were adsorbed onto soil particles. Accumulation seemed to be a function of microtopography in the basins and the resultant residence time of standing water.

However, a shortcoming of a dry basin's ephemeral pool is that in the "first flush" period when runoff from pavements tends to be highly polluted, residence time is short, because pool volume is small at the beginning of the event. In the middle of the event, residence time is longer because the ponded volume is still large while the outflow is receding, but by that time the first flush has been pushed out by later inflows.

The drawdown time can be prolonged to improve sediment settling and capture. In some locales drawdown time for a design storm has been set judgmentally at between 12 hours (one-half day) and 36 hours (3 days). Given a ponded volume based on local standards, the average drawdown rate required to produce a given drawdown time is yielded by

$$q_{ed}\text{avg} = 12.1 \, Q_{ed} / t_{ed}$$

where

$q_{ed}\text{avg}$ = average drawdown rate for the extended-detention volume, cfs

Q_{ed} = ponded volume for extended detention, af

t_{ed} = extended-detention drawdown time, hr

12.1 = conversion factor, cfs per af/hr

An outlet can be sized to produce the required drawdown rate and, thus, drawdown time. The rate of outflow through an orifice or weir is a function of head. During drawdown, the average head $H_{ed}\text{avg}$ can be assumed equal to ⅔ of the maximum head (Figure 8.5). The outlet can be sized to discharge q_{ed} avg at the average head $H_{ed}\text{avg}$, using the equations and charts presented in Chapter 6.

The maximum extended-detention head H_{ed} can be found by laying out the basin on a grading plan with the required pool volume. Use the contour planes method to check that you have drawn the pool volume correctly, and read the difference in elevation between the water surface and the outlet. Alternatively, a preliminary estimate of maximum head can be obtained by applying general knowledge of geometry and assumptions about local topog-

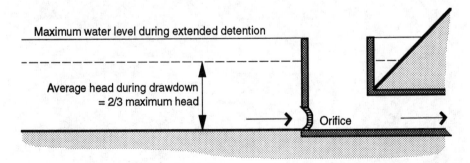

Figure 8.5 Average water level during drawdown.

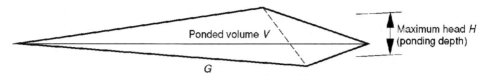

Figure 8.6   Simplified geometry of a ponded volume in a swale or valley.

raphy (Figure 8.6). If a basin formed by ponding in a swale or valley is assumed to occupy a prone right pyramid (in which volume = base area × height/3), and the side slopes of the valley are twice as steep as the hydraulic length, then the head is a function of volume and slope:

$$H_{ed} = 63.9 \, (G^2 \, Q_{ed})^{1/3}$$

where

H_{ed} = head at the deepest part of the pond, ft

G = gradient along the length of the swale where pond is located, ft/ft

Q_{ed} = ponded volume, af

Using the same geometry of a prone pyramid, the area of the extended-detention pond A_{ed} in acres can be estimated from,

$$A_{ed} = 3 \, Q_{ed} / H_{ed}$$

A dry basin can fulfill both water quality and flood control functions if it is explicitly designed for both. An outlet to control both extended detention and flood storage is a compound conveyance combining, commonly, an orifice at a low elevation and a weir at a higher elevation (Figure 8.7). The extended-detention volume is contained between the orifice and the weir. Additional flood storage is held above the weir; the depth of water over the weir is the head driving water through it. During the maximum flood control ponding, the head on the orifice is the total depth from the orifice to the flood-control water surface.

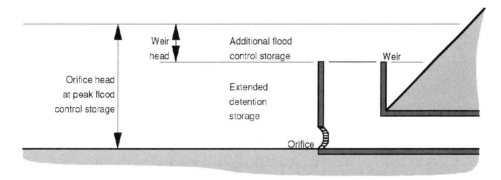

Figure 8.7   Series of elevations in a basin that combines extended detention and flood control.

With both the orifice and the weir operating, the allowable q through the weir is equal to the total allowable $q_p$out, minus the q through the orifice. Similarly, the storage volume required above the weir is equal to the total required detention volume Q_{det}, minus the extended-detention volume Q_{ed} below the weir.

The effectiveness of dry extended detention can be criticized on a number of points. Because the ponding is by definition brief, only coarse sediment particles can settle out and biochemical degradation of dissolved constituents does not have time to begin. The first flush gets little treatment. When a storm begins, the velocity of initial runoff across the basin floor can resuspend previously deposited sediment. Nevertheless, dry extended detention is practiced in some locales, and it is convenient for teaching the skills of manipulating time, flow, and volume.

Analytical Exercises

1. Using algebra, reproduce the derivation of the equations $H_{ed} = 63.9(G^2 \, Q_{ed})^{1/3}$ and $A_{ed} = 3Q_{ed}/H_{ed}$. Begin with the volume of a right pyramid (base area × height/3) and the assumption that the side slopes of the swale or valley are twice as steep as the hydraulic length.

2. Repeat the preceding derivation, with the assumption that the side slopes are 10 times as steep as the hydraulic length. How much difference does the change in steepness of side slopes make to the estimated head?

Dry Extended Detention Exercise

Exercise 8.1 adds a dry extended detention basin to the site for which you designed flood control in Exercise 7.1. The extended-detention outlet is an orifice; the flood control outlet is a rectangular weir. The SCS runoff calculations for this site were completed in Exercise 4.4.

Summary of Process

1. From Exercise 4.4, obtain drainage area A_d in acres and slope G along the hydraulic length in ft/ft.

Dry Extended Detention

2. From local standards, obtain required extended-detention volume expressed as watershed depth in inches, and minimum extended-detention drawdown time t_{ed} in hours.

3. Convert the basin's watershed equivalent depth to extended-detention volume Q_{ed} in af using $Q_{ed} = $ depth $\times A_d / 12$.

4. Find required average drawdown rate q_{ed} avg in cfs using q_{ed} avg $= 12.1 \, Q_{ed}/t_{ed}$.

5. Find maximum extended-detention head H_{ed} in ft from a grading plan or from simplified geometry of pond volume, $H_{ed} = 63.9(G^2Q_{ed})^{1/3}$.

6. Find average head during drawdown using H_{ed}avg $= 2/3 \, H_{ed}$.

7. Find orifice diameter from Figure 6.24 or the orifice equation.

8. Find extended-detention pond area A_{ed} in ac from a grading plan by using $A_{ed} = 3 \, Q_{ed} / H_{ed}$.

Flood Control Detention

9. From Exercise 7.1, obtain flood-control detention volume Q_{det} in af.

Exercise 8.1   Dry extended detention

	Site 1	Site 2
	Determining data	
Drainage area A_d (from Exercise 4.4)	= _____ ac	_____ ac
Slope G along hydraulic length (from Exercise 4.4)	= _____ ft/ft	_____ ft/ft
	Dry extended detention	
Pool volume as equivalent watershed depth (from local standards)	= _____ in	_____ in
Minimum drawdown time t_{ed} (from local standards)	= _____ hr	_____ hr
Extended-detention volume Q_{ed} = depth $\times A_d$ / 12	= _____ af	_____ af
Average drawdown rate q_{ed}avg = 12.1 Q_{ed} / t_{ed}	= _____ cfs	_____ cfs
Maximum head H_{ed} = 63.9$(G^2Q_{ed})^{1/3}$	= _____ ft	_____ ft
Average head H_{ed} avg = (2/3) H_{ed}	= _____ ft	_____ ft
Outlet orifice diameter (from Figure 6.24)	= _____ in.	_____ in.
Pond area A_{ed} = 3 Q_{ed} / H_{ed}	= _____ ac	_____ ac
	Flood control detention	
Flood detention volume Q_{det} (from Exercise 7.1)	= _____ af	_____ af
Flood storage above weir = $Q_{det} - Q_{ed}$	= _____ af	_____ af
Head above weir H_{det} = $(Q_{det} - Q_{ed})$ / A_{ed}	= _____ ft	_____ ft
Total head above orifice = $H_{ed} + H_{det}$	_____ ft	_____ ft
Discharge through orifice q_{ed} at total head (from Figure 6.24)	= _____ cfs	_____ cfs
Total outflow rate $q_p$out (from Exercise 7.1)	= _____ cfs	_____ cfs
Discharge over weir = $q_p$out - q_{ed}	= _____ cfs	_____ cfs
Length L of flood-control weir (from Figure 6.17)	= _____ ft	_____ ft

10. Find portion of flood storage volume above flood-control weir using $Q_{det} - Q_{ed}$.
11. Estimate head above weir H_{det} in ft using $H_{det} = (Q_{det} - Q_{ed})$ / Aed.
12. Find total head above orifice using $H_{ed} + H_{det}$.
13. Find discharge through orifice q_{ed} in cfs at total orifice head from Figure 6.24.

14. From Exercise 7.1, obtain total allowable outflow rate $q_p$out in cfs.
15. Find allowable rate of flow over weir from $q_p$out $- q_{ed}$.
16. Find length of weir L using the weir equation or Figure 6.17.

Discussion of Results

1. What is the approximate proportion of your site that must be occupied by its extended-detention pool (A_{ed} / A_d)? In your judgment, does the basin occupy too much land to be reasonable purely in terms of land use?
2. What portion of the total peak outflow (cfs/cfs) goes through the orifice, and how much over the weir? In some sites, no flow is allowed over the weir at all during the design storm; in fact, the orifice may have to be further constricted to bring down outflow rate to that needed for flood control. In these cases, the weir can be the emergency outflow, to go into operation when the flood-control design storm is exceeded.

WET BASINS

A permanent pool has important—almost essential—advantages for extended detention. By adding volume to the ponded storage before a storm begins, a permanent pool lengthens residence time, most importantly in the "first-flush" period when residence time is most needed. It continues settling and biodegradation long after a storm is over. A greater proportion of the suspended sediment settles out of the runoff; the low velocity in the pool protects previously settled sediment from resuspension. As residence time extends beyond 14 days, biochemical decomposition of dissolved and adsorbed constituents becomes significant; algae and microorganisms degrade and remove constituents such as nutrients, metals, and organic chemicals.

The effect of long-term ponding in a permanent pool can be evaluated with the monthly water balance. With this approach, the design flow is the largest average monthly runoff. If the pond works during the month with the most runoff, then on the average it will work for the other months as well. The treatment capacity of the pond would be exceeded only in the relatively rare periods with throughflow greater than that in the month with the largest average runoff.

Figure 8.8 shows a permanent pool at the Orange County Civic and Convention Center near Orlando, Florida. The weir in the foreground maintains the stream's elevation as it discharges from a five-acre cypress wetland preserve. The pool in the background receives runoff from the site's impervious roofs and pavements. In the distance are fountains that aerate the water while enlivening the entry area of the convention center. On the shore opposite the building is a shallow zone with native wetland plants. The space above the permanent pool's surface, up to the weir elevation, is in reserve to detain runoff volumes when large storms occur.

Pool Residence Time and Volume

Heinemann (1981) established a relationship between sediment trap efficiency and long-term average residence time. It is given by (derived from Heinemann, 1981):

$$t_r = (96.4 + 4.38\,E) / (97.16 - 1.02\,E)$$

Figure 8.8 Permanent pool at the Orange County Civic and Convention Center, Florida.

where average residence time t_r is in days and sediment trap efficiency E is in percentage. This relationship applies to the trapping of overall mixtures of alluvial sediment. It applies to permanent pools with drainage areas smaller than 9,600 acres, and from which water discharges from the surface of the pool.

Figure 8.9 shows the results of Heinemann's equation. Low trap efficiencies result from very short residence times during which only coarse particles settle out. But increasing residence times, up to about seven days, rapidly produces higher trap efficiencies as smaller particles begin to be affected. At higher levels of trap efficiency, required residence time increases rapidly as the finest particles stay in suspension longer. It is hard to get sediment trap efficiency higher than 80 percent even in the most favorable ponds. To use the graph in design, enter from the bottom at the desired trap efficiency, move up to the curve, and thence to the left to read the required permanent pool residence time.

The pool volume required to create a certain average residence time is easily derived,

$$Q_p = t_r\, q$$

where

$$Q_p = \text{volume of permanent pool, af}$$

$$t_r = \text{residence time, days}$$

$$q = \text{inflow, af/day}$$

The value of q should be total flow $q\Sigma$ (the sum of direct runoff and base flow) where the pool is located in a stream with perennial or seasonal flow that continues between

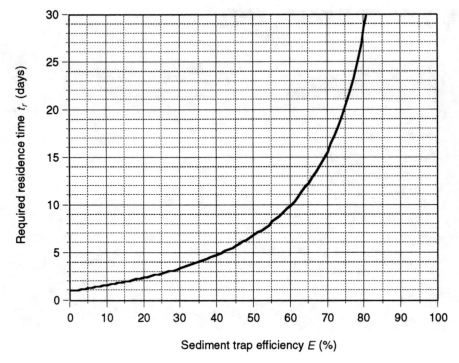

Figure 8.9 Average residence time required to achieve overall sediment trap efficiency, in permanent pools with surface discharge (derived from Heinemann, 1981).

storms. It should be Q_d, direct runoff only, where a pool is located with such a small drainage area that it receives inflow only during storms and base flow bypasses the pool through the subsurface, discharging to springs at lower elevations. Apply to the above equation the largest monthly value of Q_d or $q\Sigma$ found in the drainage area's water balance.

Trap efficiency for chemical constituents such as phosphorus and metals is greater than that for sediment, because chemical constituents are adsorbed almost exclusively onto the clay particles, a large portion of which stay in suspension and are discharged from the pond after the coarse particles have settled out.

Drawdown time during a specific design storm need not be provided for where a permanent pool is designed for long-term performance with total monthly runoff. The volume of the permanent pool itself produces the required residence time, both during and between storms. Nevertheless, in some locales a specially constricted extended-detention outlet is required to prolong drawdown, even from a permanent pool. The outlet for such a provision would establish the permanent pool level and control a designated drawdown volume, as it does in a dry basin.

The following sections describe landscape design criteria for assuring that a permanent pool of given Q_p is, in fact, effective at improving water quality and other aspects of the environment.

The "Live Zone"

The volume of water that outlets from a pond during a storm is equal to the volume that flows in. Newly inflowing water pushes some "old" water ahead of it, through the pond and

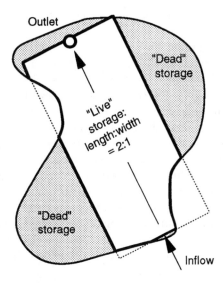

Figure 8.10   "Live" storage in a 2:1 rectangle and "dead" storage outside it.

out. While the water level rises and falls, individual drops of water within the pond are grad-ually replaced.

A permanent pool's "live zone" is the portion of the volume on and near the direct flow line between the pond's inlet and outlet, where the replacement of "old" water with "new" takes place. The live zone is commonly assumed to lie within a rectangle with the length (the distance between inflow and outlet) at least twice the width, as shown in Figure 8.10. Areas outside the rectangle are "dead" storage, in the sense that they do not participate in the treatment of throughflow. In evaluating pool volume Q_p for active water quality treat-ment, only the live area should be counted.

A pond's length:width ratio can be increased to 2:1 or more by relocating inlets and out-lets, creating peninsulas or islands to lengthen the flow path, or lengthening the overall pond with earthwork. Shaping a pond into a geometric rectangle to get completely live storage is not necessary, as long as at least part of the pond contains a live storage area as big as the required Q_p. To reshape the edges to eliminate dead storage, when not necessary, would be an expensive and ecologically crude approach to "efficiency."

Sediment Storage

Beneath the required volume of water in a pond, an additional volume must be set aside for storage of sediment that has settled out of the water. Providing a large sediment storage vol-ume allows the pond to operate for a long time before sediment has to be removed to restore the pond's capacity. On the other hand, constructing a large volume may require a large expense. In typical urban watersheds with stable land use, a sediment storage volume equal to ½ inch of watershed runoff may allow sediment removal less than once every 20 years. Of course, the cycle would have to be much shorter where there is an active source of eroding sediment in the watershed; when that happens, it is imperative to stabilize the sed-iment source in addition to restoring the capacity of the pond.

Given the equivalent watershed depth of sediment storage D_s in inches, the sediment vol-ume in af can be found by using

$$\text{Sediment volume} = D_s \, A_d \, / \, 12$$

Each time sediment is removed, the removal costs about 20 to 40 percent of the cost of basin construction (Schueler and Helfrich, 1989). The removal cost can be reduced by anticipating the procedure in design. Installing a drain pipe in the bottom of the pond outlet allows earth-moving equipment to enter the drained pond for sediment removal, which is much cheaper than using a dragline to reach in from the shore. Installing the outlet riser in the embankment, not in the pool, allows easy access for opening the drain. Dedicating an access easement to the pond from a public street allows maintenance access without trouble.

Sediment generated during the construction period can be trapped in an extended-detention pond, as can any other sediment. The sediment storage volume must be adequate to hold the amount of sediment expected to be generated during the construction period. The accumulated sediment should be removed when construction is over and the drainage area is stabilized, so as to allow the pond to commence its long-term operation with a stable watershed. In some areas it is a rule of thumb to expect 67 cubic yards of sediment for every acre of construction area during an entire construction period; this is equivalent to one half inch of erosion over the construction area.

Pool Depth and Area

As shown in Figure 8.11, pool depth is measured from the surface of the permanent pool to the top of the sediment storage zone. Settling has effectively occurred when a settling particle reaches the bottom of the pool and enters the sediment storage zone.

Spreading out a given pool volume Q_p over a large area makes settling more effective by reducing depth to the sediment storage zone and, thus, settling time. For a given mineral particle size having a given settling velocity, the average pool depth required to reduce settling time below residence time is yielded by

$$D_p\text{avg} = V_s \, t_r$$

where

$D_p\text{avg}$ = average depth of permanent pool in the "live" zone, ft

V_s = settling velocity for a selected particle size, ft/day

t_r = residence time in the permanent pool, days

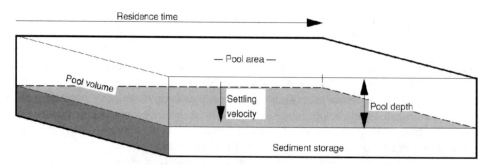

Figure 8.11 Relationship between settling and depth in a given pool volume.

For a given pool depth, pool area can be estimated from

$$A_p = Q_p / D_p \text{avg}$$

where

$$A_p = \text{pool area, ac}$$

$$Q_p = \text{pool volume, af}$$

In addition to supporting the effectiveness of physical settling, a large pool area exposes the water to air and light, supporting microbiotic activity for uptake and decomposition of dissolved and adsorbed constituents.

The dimensions in feet of a 2:1 live-zone rectangle that a pool would occupy can be derived from the geometry of the rectangle:

$$\text{Length} = 295 \, A_p^{1/2}$$

$$\text{Width} = \text{length} / 2$$

Within a pool of a given average depth, zones of different depths, greater and less than the average, can perform different detailed functions in water quality improvement and environmental enhancement.

One type of deep zone is a forebay, located where runoff discharges into the basin. The abrupt slowing down of the runoff when it reaches the pool causes sand to be quickly deposited. A forebay can be shaped to spread out runoff laterally, discharging it uniformly into a downstream wetland. The water's energy is dissipated, so it cannot erode shallow zones when it gets to them downstream.

Another type of deep zone is an outlet pool at the downstream end of a basin, which keeps an outlet pipe or trash rack submerged so that the basin retains floating debris and oil.

Zones of water shallower than about 18 inches are habitats for rooting of aquatic vegetation. These are called littoral zones or wetland zones. On the plentiful surfaces of aquatic plants and in the light and oxygen near the water surface, microbiotic activity degrades dissolved pollutants when residence time is sufficient. Plants stabilize settled sediment by reducing water velocity and by binding with their roots. Shallow, gently sloping zones around basin edges provide the human benefit of safety for those who stumble into the basin. However, shallow areas should be used with caution in exposed parts of wide ponds where wind-generated turbulence could resuspend bottom sediment.

Flood-Control Detention

Additional reservoir volume above the permanent pool can act as detention storage for flood control. The permanent pool is not counted in temporary flood storage. Thus, the flood storage volume Q_{det} obtained using the procedures presented in Chapter 7 must be contained entirely above the permanent pool; the surface of the pool is the bottom of the flood storage volume.

The head H_{det} driving water through the flood-control outlet can be found by laying out the basin, including both the permanent pool and the flood storage volume, on a grading plan and reading the difference in elevation from the permanent pool surface to the top of the flood volume. Alternatively, a preliminary estimate of H_{det} can be obtained by assuming

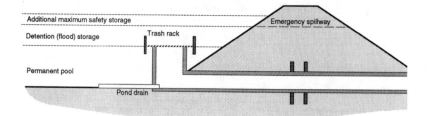

Figure 8.12 The outlet of a basin that combines flood detention with a permanent pool for extended detention.

that the horizontal area covered by the detention is equal to that of the permanent pool, A_p. In this case, $H_{det} = Q_{det} / A_p$.

Given the detention head H_{det} and outflow rate q_p out from the procedures described in Chapter 7, the detention outlet's type and size can be selected from the charts in Chapter 6.

With flood control, the outlet from a permanent pool combines several features that must be stacked up in elevation, as shown in Figure 8.12. The flood-control outlet sets the level of the permanent pool. A trash rack, or other provision for outletting only water from below the pool surface, retains floating debris and oil in the pool. The emergency spillway safely discharges the largest possible flows, at a head above the flood-control storage.

Wetland Plantings

Wetland plants such as rush (*Juncus* sp.), bulrush (*Scirpus* sp.), reed (*Phragmites* sp.), cattail (*Typha* sp.), and submerged aquatics can be planted in shallow areas. Throughflowing stormwater molds the plants' habitats and constrains the types and amounts of plants that will do well. Flows from an urban drainage area are "flashy," with low base flows and abruptly high peaks: the environment is stressed. Invasive "weeds" such as *Typha* and *Phragmites* can be common, because they thrive in disturbed environments. A pond's potential vegetative diversity depends on the complexity of physical habitats (flow, depth, light, nutrient and sediment loads) the plants are given to live in and the availability of seed sources to take advantage of those habitats.

Densely rooted plants slow water down and distribute its flow uniformly. They stabilize the sediment they are rooted in, preventing it from being resuspended.

Dense stems and leaves provide surface area where biochemically active algae and microorganisms live. After microbiota have made nutrients available in the sediment, plants can take up the chemicals through their roots. Stands of mixed species, having various adaptive mechanisms and life cycles, may remove overall constituents most effectively. However, tall emergent species such as cattail have been controversial because they may deprive pond water of oxygen at times during their life cycles by contributing only dead, decaying vegetative matter to the pond bottom. Periodic harvesting of vegetation can slow the accumulation of chemical constituents in a pond.

Overall Configurations

When all the criteria for permanent pools are applied together on given sites, the ponds can fall into patterns of overall configurations.

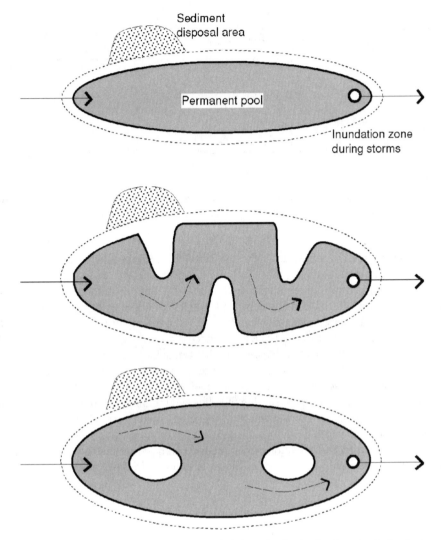

Figure 8.13 Schematic configurations of extended-detention basins in plan: simple (*top*), sinuous (*middle*) and broken (*bottom*).

Configurations in plan are shown in Figure 8.13. Each of the drawings shows a permanent pool with inlet and outlet, the surrounding inundation zone where water rises during storms, and, near the inlet, an area reserved for future sediment disposal when removal is necessary. All parts of the permanent pools shown in the figure can be counted in a properly proportioned "live" storage volume. The simple plan has an open pond with little "edge"; flow length is a function only of the overall dimension of the pond. The sinuous plan is formed of bays and peninsulas, creating a longer, meandering flow path. The broken plan has islands diverting the flow path, which becomes longer as it braids through the remaining channels. As compared with a simple plan, sinuous and broken plans have longer flow paths and, hence, higher treatment performance, more edge for ecological interactions, and greater diversity of habitats for all types of organisms. The islands in a

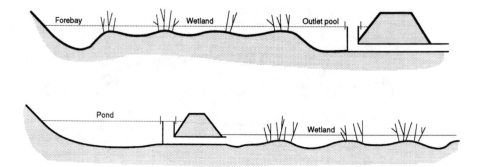

Figure 8.14 Schematic configurations of extended-detention basins in profile: extended-detention wetland (*top*) and pond-wetland system (*bottom*) (after Schueler, 1992, Figure 5).

broken plan can be of benefit specifically to waterfowl, which use islands as protected resting and nesting places.

Configurations in profile are shown in Figure 8.14. The illustrations show the earth forms that produce different ponding depths and suggest the diversity of vegetation that may respond to them. Either configuration can be laid out along any of the kinds of flow paths shown in Figure 8.13. An extended-detention wetland may incorporate a forebay pool near the inlet, to slow the flow and disperse it into the wetland, and an outlet pool to submerge the primary outlet and retain floating debris. In between, water flows through diverse shallow zones where it comes into contact with vegetation and soil. In contrast, a pond-wetland system begins with a pond for the settling of solids and trapping of debris. The discharging water flows through a shallow wetland for "polishing" by removal of finer sediment and contact with soil and vegetation.

Installation of extended detention basins can have both positive and negative impacts on the ecology of a site and a watershed, as shown in Figure 8.15. Installing a basin produces

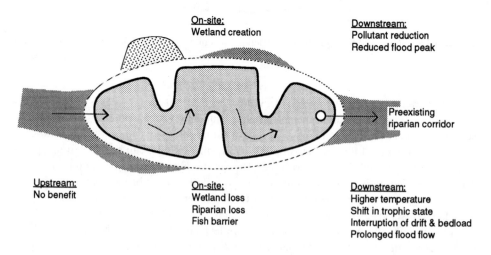

Figure 8.15 Impacts of installation of an extended-detention basin: positive (*top*) and negative (*bottom*) (after Schueler, 1992, Figure 18).

a new artificial wetland, with the ecological benefits ordinarily provided by wetlands, and contributes to pollution reduction and flood control downstream. On the other hand, the basin is installed in a preexisting fluvial and riparian system. Its installation destroys the wetlands and riparian zone that may have previously existed there and imposes a barrier to fish migration. Downstream, the impoundment raises stream temperature, shifts the trophic state of the aquatic ecosystem, alters flows of drift and sediment, and prolongs flood flows when they occur. Water quality upstream of the impoundment remains unchanged. The positive and negative impacts have to be weighed against each other and used to select locations and devise specific designs within the context of the objectives and needs of each specific project.

Under certain circumstances, natural wetlands can improve the quality of water that passes through them. They act as natural extended-detention basins. However, in order not to damage the stability and value of existing wetlands, only small quantities of runoff should be discharged into them per area of wetland; they must be protected from overloads of sediment during construction; and stormwater should enter at a number of points via sheet-flow and made to distribute throughout the area of the wetland.

Analytical Exercise

Figure 8.9 showed that it is hard to obtain sediment trap efficiency higher than 80 percent. What type of geographic location, and which characteristics of basin and watershed, would help to produce higher sediment trap efficiencies? Name some specific locations that may exemplify some of these conditions. What characteristics of Lake Ellyn and its watershed could have produced its high trap efficiency?

Wet Extended-Detention Exercise

Exercise 8.2 adds a permanent pool for extended detention to the flood-control reservoir you designed in Exercise 7.1. A standpipe acting as a circular weir sets the permanent pool level and controls rate of outflow during flood-control design storms. The basin's inflow is total flow, the sum of direct runoff and base flow; the calculation of this amount was completed in the water-balance exercises in Chapter 5. The SCS estimation of storm runoff for this site was completed in Exercise 4.4.

Summary of Process

1. From Exercise 4.4 obtain drainage area A_d in acres.

   Sediment Storage

2. From local standards obtain the required sediment storage volume D_s, expressed in watershed inches. Convert the amount to af with

$$D_s \text{ (af)} = D_s \text{ (inches)} \times A_d / 12$$

   Permanent Pool

3. From local standards obtain the required sediment trap efficiency E.

4. Obtain required residence time t_r in days from Heinemann's equation or Figure 8.9.

5. From Exercise 5.5 obtain the largest monthly total flow $q\Sigma$ in inches after development. Convert $q\Sigma$ to af/day with

Exercise 8.2 Wet extended detention

	Site 1	Site 2

Determining data

Drainage area A_d
  (from Exercise 4.4) = _____ ac _____ ac

Sediment storage

Sediment storage watershed depth
  D_s (from local standards) = _____ in _____ in
Sediment storage volume
  $= D_s A_d / 12$ = _____ af _____ af

Permanent pool

Required sediment trap efficiency
  E (from local standards) = _____ % _____ %
Required average residence time t_r
  (from Figure 8.9) = _____ days _____ days
Largest monthly total flow $q\Sigma$ after
  development (from Exercise 5.5) = _____ in./mo _____ in./mo
Largest monthly total flow $q\Sigma$ in
  af/day $= q\Sigma$ in in/mo $\times A_d / 365$ = _____ af/day _____ af/day
Required permanent pool volume
  $Q_p = t_r q\Sigma$ = _____ af _____ af
Design settling velocity V_s
  (from local standards) = _____ ft/day _____ ft/day
Average pool depth $D_p$avg
  $= V_s t_r$ = _____ ft _____ ft
Pool area A_p
  $= Q_p / D_p$avg = _____ ac _____ ac
Pool length
  $= 295 A_p^{1/2}$ = _____ ft _____ ft
Pool width
  $=$ length $/ 2$ = _____ ft _____ ft

Flood-control outlet

Flood detention volume Q_{det}
  (from Exercise 7.1) = _____ af _____ af
Peak outflow $q_p$out
  (from Exercise 7.1) = _____ cfs _____ cfs
Head H_{det} over flood control outlet
  $= Q_{det} / A_p$ = _____ ft _____ ft
Possible rectangular weir outlet
  length (from Figure 6.17) = _____ ft _____ ft
Possible circular weir outlet
  diameter (from Figure 6.19) = _____ in. _____ in.
Possible vee-notch weir outlet
  angle (from Figure 6.22) = _____ ° _____ °

Total reservoir volume

Total pond volume
  $= D_s + Q_p + Q_{det}$ = _____ af _____ af

$$q\Sigma \ (\text{af/day}) = q\Sigma \ (\text{in/mo}) \times A_d \ / \ 365$$

6. Compute required permanent pool volume Q_p in af using $Q_p = t_r \, q\Sigma$.
7. Obtain design settling velocity V_s in ft/day from local standards.
8. Find required average pool depth $D_p$avg in ft from $D_p$avg $= V_s \, t_r$.
9. Find pool area A_p in acres using $A_p = Q_p \ / \ D_p$avg.
10. Find dimensions in ft of the "live" 2:1 pool area using length = 295 $A_p^{1/2}$ and width = length / 2.

Flood-Control Outlet

11. From Exercise 7.1 obtain flood detention volume Q_{det} in af and peak flow allowed through outlet $q_p$out in cfs.
12. Find head over the outlet H_{det} in ft from $H_{det} = Q_{det} \ / \ A_p$.
13. For each possible outlet configuration, find the size for the given H_{det} and $q_p$out from the appropriate equation or chart in Chapter 6.

Total Reservoir Volume

14. Find total reservoir volume as the sum of Q_p, Q_{det}, and the sediment storage volume.

Discussion of Results

1. Which required the larger basin volume and area on your site, a permanent pool for total monthly runoff (Exercise 8.2) or a dry basin for a passing storm (8.1)? Which one do you think will be a more effective trap of sediment and dissolved constituents?
2. Which site requires the larger permanent pool? What factors contributed to this requirement?
3. What is the approximate proportion of your site that must be occupied by its pool ($A_p \ / \ A_d$)? In your judgment, does the basin occupy too much land to be reasonable purely in terms of land use?
4. If your response to the second part of question 3 was yes, some approaches you might take to solving the problem include choosing another site for this type of development, reducing the trap efficiency to be attained, and selecting some other type of hydrologic process to improve water quality. Which approach would you most likely choose? Why? What specific site features or changes to your design would you have to build in order to implement your approach?

SUMMARY AND COMMENTARY

Extended detention is another modification of conveyance. Like flood-control detention, it modifies the rate of surface flow. By modifying rate adequately and in the right environment, it modifies the quality of throughflowing water.

Quantitative certainty of downstream water quality results should not be expected (Griffin, 1995). Every extended-detention basin is only partly effective, and its degree of effectiveness is influenced by numerous site-specific factors. In a watershed that contains some preexisting urban land use, effective improvement of water quality depends on retrofitting at least some of the older developed areas, as well as application in new developments.

Like flood-control detention, extended detention does not affect storm flow volume. The entire volume eventually outlets from a basin and continues running downstream. Extended detention, alone, does not significantly address groundwater or stream base flow.

An extended-detention basin is a facility in a site development program. It must be given adequate space in a site plan, as well as a constructive landscape design.

The results of hydrologic calculations for extended detention give the required volume of the basin and constraints on its length, width, and depth. These few results are parameters for site design. As long as you meet these few criteria you are free to design the site in any way that is needed for the environment and the people who live within it.

REFERENCES

Friedman, J., 1985, *Wetlands Hydrology and Sedimentation, Implications for the Design and Management of Wetland Preserves*, Seattle, Wash.: The Nature Conservancy.

Griffin, Carol B., 1995, Uncertainty Analysis of BMP Effectiveness for Controlling Nitrogen from Urban Nonpoint Sources, *Water Resources Bulletin* vol. 31, no. 6, pp. 1041–1050.

Grizzard, T. J., C. W. Randall, B. L. Weand, and K. L. Ellis, 1986, Effectiveness of Extended Detention Ponds, *Urban Runoff Quality: Impact and Quality Enhancement Technology*, New York: American Society of Civil Engineers.

Hammer, Donald A., editor, 1989, *Constructed Wetlands for Wastewater Treatment*, Chelsea, Mich.: Lewis Publishers.

Hartigan, John P., 1989, Basis for Design of Wet Detention Basin BMPs, *Design of Urban Runoff Quality Controls*, pp. 122–143, New York: American Society of Civil Engineers.

Heinemann, H. G., 1981, A New Sediment Trap Efficiency Curve for Small Reservoirs, *Water Resources Bulletin* vol. 17, no. 5, pp. 825–830.

Kadlec, Robert H., and Robert L., Knight, 1995, *Treatment Wetlands*, Boca Raton: Lewis.

Kusler, Jon A., and Mary E. Kentula, editors, 1989, *Wetland Creation and Restoration: The Status of the Science* (reprinted from EPA/600/3-89/089a), Washington: Island Press.

LaBaugh, James W., 1986, Wetland Ecosystem Studies from a Hydrologic Perspective, *Water Resources Bulletin* vol. 22, no. 1, pp. 1–10.

Livingston, Eric H., 1989, The Use of Wetlands for Urban Stormwater Management, *Design of Urban Runoff Quality Controls*, pp. 467–489, New York: American Society of Civil Engineers.

Marble, Anne D., 1992, *A Guide to Wetland Functional Design*, Chelsea, Mich.: Lewis Publishers.

Maryland Water Management Administration, 1987, *Design Procedures for Stormwater Management Extended Detention Structures*, Baltimore: Maryland Department of the Environment, Water Management Administration.

Nix, Stephan J., 1985, Residence Time in Stormwater Basins, *Journal of Environmental Engineering* (American Society of Civil Engineers), vol. 111, no. 1, pp. 95–100.

Roesner, Larry A., Ben Urbonas, and Michael B. Sonnen, editors, 1989, *Design of Urban Runoff Quality Controls*, New York: American Society of Civil Engineers.

Schueler, Thomas R., 1993, *Design of Stormwater Wetland Systems: Guidelines for Creating Diverse and Effective Stormwater Wetlands in the Mid-Atlantic Region*, Washington: Metropolitan Washington Council of Governments.

Schueler, Thomas R., and Mike Helfrich, 1989, Design of Extended Detention Wet Pond Systems, *Design of Urban Runoff Quality Controls*, pp. 180-202, New York: American Society of Civil Engineers.

Striegl, Robert G., 1987, Suspended Sediment and Metals Removal from Urban Runoff by a Small Lake, *Water Resources Bulletin* vol. 23, no. 6, pp. 985–996.

U.S. Environmental Protection Agency, 1990, National Pollutant Discharge Elimination System Permit Application Regulations for Storm Water Discharges, Final Rule, *Federal Register* vol. 55, no. 222, pp. 47990–48091.

U.S. Soil Conservation Service, 1992, *Wetland Restoration, Enhancement or Creation*, National Engineering Handbook, Section 13, Washington: U.S. Soil Conservation Service.

Van der Leeden, Frits, Fred L. Troise, and David Keith Todd, 1990, *The Water Encyclopedia*, second edition, Chelsea, Mich.: Lewis Publishers.

Walker, W. W., 1987, Phosphorus Removal by Urban Runoff Detention Basins, of *Lake and Reservoir Management,* vol 3, pp. 314–326, Washington: North American Lake Management Society.

Whipple, William, Jr., and Joseph V. Hunter, 1981, Settleability of Urban Runoff Pollution, *Journal Water Pollution Control Federation* vol. 53, no. 12, pp. 1726–1731.

Wigginton, Parker J., Clifford W. Randall, and Thomas J. Grizzard, 1983, Accumulation of Selected Trace Metals in Soils of Urban Runoff Detention Basins, *Water Resources Bulletin* vol. 19, no. 5, pp. 709–718.

Wulliman, James T., Mark Maxwell, William E. Wenk, and Ben Urbonas, Multiple Treatment System for Phosphorus Removal, *Design of Urban Runoff Quality Controls,* pp. 239–257, New York: American Society of Civil Engineers.

CHAPTER 9

INFILTRATION

Infiltration is the soaking of water into the ground. It involves bringing water into prolonged contact with soil at any possible opportunity in a site, especially in closed basins from which the primary outlet is into the soil.

Infiltration basins are depressions in the earth that capture and hold direct runoff while it infiltrates. Infiltration basins have been used for urban stormwater since about 1930. There are more than 20,000 of them in operation in the United States.

Conveyance and detention perpetuate surface types of flows even while managing flow rate and quality. Infiltration is qualitatively different from conveyance and detention because it puts direct runoff into a new kind of place, where it undergoes new kinds of processes. Urban infiltration constitutes the restoration of a site's hydrologic process. It restores groundwater to the earth and balanced flow regimes to streams. In addition to addressing flooding and erosion, which are targeted by conveyance and detention systems, infiltration supports groundwater recharge, stream base flows, water quality, aquatic life, and water supplies. Because it turns the hazard of storm flows into the resource of base flows, it is environmentally the most complete solution to the problem of urban stormwater. You should try to infiltrate as much as you can; turn to other approaches only to treat the remaining runoff that cannot be infiltrated. Water belongs in the soil; returning it there is a basic task for urban design.

INFILTRATION'S CONTEXT IN LANDSCAPE HYDROLOGY

A review of how hydrology works in nature and how urban development changes it will make clear the importance of infiltration to the health of the landscape and the people who live within it.

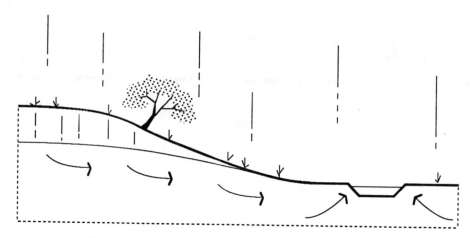

Figure 9.1 Landscape hydrology without impervious surfaces.

Figure 9.1 shows how, under natural conditions, a large portion of rainfall tends to infiltrate the soil and join the soil moisture. In densely vegetated humid regions most rainwater infiltrates where it falls because plants, leaf litter, and soil organic matter break up raindrops' energy and let water soak into numerous soil voids. In sparsely vegetated arid regions natural infiltration also operates, except that rainwater sometimes has to travel some distance down drainage channels before it has time to infiltrate.

A site's ecosystem uses some of the infiltrated water to build and maintain itself through evapotranspiration. The rest of the water joins the site's groundwater in the voids of soil and rock.

Subsurface storage is large. Water may reside in the subsurface for weeks or months before reaching streams. Water emerges from subsurface storage so gradually that its discharge is part of the streams' base flow, not the original storm flow. The emerging flow has become a different type of flow, supporting different types of resources.

Figure 9.2 shows that, with development, impervious surfaces deflect rainfall from its natural course. They send surface runoff directly to streams, making floods larger. Runoff moves quickly down stream systems; as a resource it is lost to the watersheds where it orig-

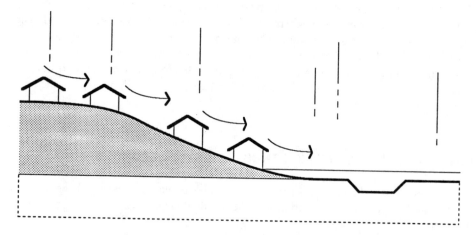

Figure 9.2 Landscape hydrology with impervious cover.

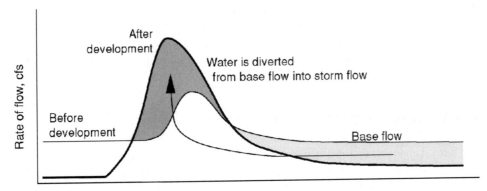

Figure 9.3   Enlargement of storm hydrograph following development with impervious surfaces.

inated. Groundwater, soil moisture and stream base flows decline. Impervious surfaces prevent water from following the natural processes of filtration, storage and gradual discharge that moderate hydrologic flows. By doing so, they aggravate natural stream flow fluctuation, making high flows higher and low flows lower. They both aggravate the hazards associated with storm flows and injure the resources supported by base flows.

Following development, the rainfall is the same as before, but there is more water in the enlarged storm hydrograph. The "extra" water was there on the site all along. Under natural conditions it infiltrated the soil and emerged long after storms were over. Impervious surfaces deflect that water into storm flow. Piling up all that volume within a compressed time, as shown in Figure 9.3, necessarily creates a higher peak flow rate. The large after-development storm hydrograph shows both the high peak rate of flow during the storm and the volume of flow that has been diverted away from base flow between storms.

Stormwater management must address base flows as well as peak flows. It must address long-term storage in subsurface soil voids as well as visible flows in surface channels. It must reinitiate the self-sustaining kinds of long-term environmental processes that occurred before impervious surfaces were installed.

Stormwater infiltration (Figure 9.4) returns surface flows to the subsurface, keeping aggravated storm flows out of streams and returning them to their native place in base

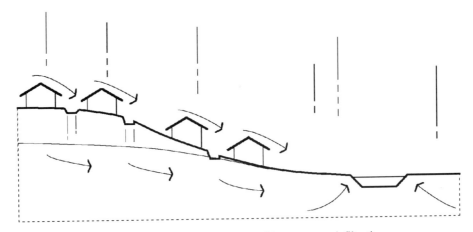

Figure 9.4   Landscape hydrology with stormwater infiltration.

flows. It takes advantage of the land's natural capacities for filtration and long-term storage. Unlike detention, infiltration can never aggravate flooding downstream, no matter how a watershed's tributaries combine, because infiltration reduces the volume of stormwater at its source. The subsurface is a resource that is waiting to be taken advantage of on every site, as nature always used it in the past.

INFILTRATION'S HYDROLOGIC EFFECT

The longest and most thoroughly documented experience with urban stormwater infiltration is that of Long Island, New York, where stormwater "recharge" started at about 1930. The objectives were to minimize storm sewer expense by eliminating runoff near the source and to replenish the island's groundwater, which is the only source for municipal supplies. As of 1990 there were more than 4,000 basins in operation, with more than a million people living in their drainage areas. Of the 45 inches of rain per year that falls on the drainage areas, the equivalent of about half, or 22 inches, infiltrates through the recharge basins.

The effect of the basins is shown by a comparison of water supply and demand (Figure 9.5). Municipal agencies draw up 150 gallons of groundwater per person per day to support the residents who live in the basins' drainage areas. Through the recharge basins, runoff puts the equivalent of 240 gallons per person per day back into the aquifer, more than replacing the withdrawals.

The excess recharge through Long Island's basins ends up subsidizing the water balance and water supply of the parts of the island without infiltration (Ku, Hegelin and Buxton, 1992). In the old and low-lying urban areas where infiltration was not implemented, recharge is about 10 percent less than it was before development because runoff is routed on the surface to streams and tidewater. In these areas the water table has fallen by as much as 3 feet; streams such as East Meadow Brook have gone dry; water supplies have disappeared. In the parts of the island where runoff is infiltrated, annual groundwater recharge is about 12 percent greater than it was before urbanization. In these areas the water table has risen by as much as 5 feet; streams flow year-round; municipal water supplies are secure. Excess groundwater flows into the areas with shortages, replenishes the water supplies of the island as a whole, and pushes contaminating brackish water out of the aquifer as it flows out to tidewater.

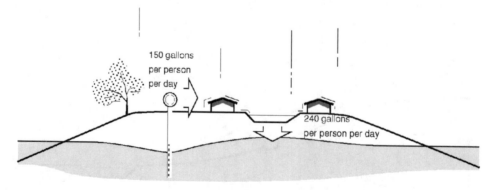

Figure 9.5   Effect of infiltration on Long Island, New York.

Infiltration has saved Long Island's people and environment. As long as Long Island's future developments include similar basins, urban areas can continue to grow without exceeding the island's native water supply.

Long Island is a clear case study because its continuous water table is a lucid barometer of net hydrologic effects and, as an island, it is markedly dependent on rainfall for its water supply. Essentially all urban landscapes can follow this example, even those with less obvious hydrologic connections. All landscapes and populations rely in one way or another on subsurface storage and filtration for treatment, replenishment, and mediation of water and its effects.

INFILTRATION'S WATER QUALITY EFFECT

Where the soil through which water infiltrates contains any degree of clay or humus, the soil is a powerful filter and dynamic ecosystem that protects streams and aquifers from urban contamination. It takes only a few inches of soil to trap and accumulate oils, metals, and nutrients. Nitrogen compounds tend to decompose and return to the atmosphere. As long as infiltrating runoff contains only the common, mostly biodegradable, constituents from residential, commercial, and office sites, and not foreign chemicals discharged directly from industry, then it is within most soils' treatment capacity (Pitt et al, 1996).

Long Island's underlying aquifer has not been measurably polluted by stormwater infiltration basins, either chemically or microbiologically (Ku and Simmons, 1986). In fact, where the aquifer was already polluted with nitrogen from other sources such as septic tanks, some recharge basins have been found to dilute nitrogen in the groundwater with fresh water,

Similar studies in the Central Valley of California (Nightingale, 1987a and b) found the quality of groundwater directly below operating infiltration basins indistinguishable from that elsewhere in the regional aquifer. Although the basins were as much as about 20 years old, typical runoff contaminants were still being trapped in the first few inches of soil below basin floors.

To ensure that the soil has adequate treatment capacity, the Washington State Department of Ecology requires infiltration basins to be lined with an "aquitard" of medium-textured soil where excessively permeable gravels make a direct conduit to the water table. Then hydrologic design must be redone to take into account this artificially limiting layer.

The accumulation of constituents in an effective infiltration basin implies that the day will come when the soil is "full" and the concentrations are toxic. In moderate soils with ordinary urban pollutant loadings, accumulation to this stage will take about 200 years. No agency has yet found it necessary to remove soil from existing infiltration basins as a result of toxic concentration. Ultimately, the only way to prevent pollutants from accumulating somewhere in the environment is to stop generating them in the first place—the pollution problem is, at source, a problem of our way of life.

THE SOIL SURFACE WHERE INFILTRATION HAPPENS

Figure 9.6 shows two directions in which the soil surface in an infiltration basin can evolve, depending on the initial ponding time immediately after construction.

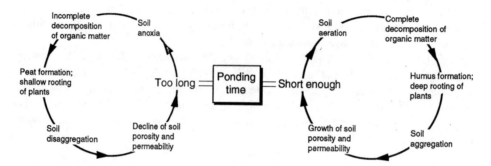

Figure 9.6 Alternative cycles in the evolution of an infiltration basin's surface.

In a basin with a short initial ponding time, the basin's soil is aerated and it supports deeply rooted vegetation. In the aerobic soil habitat, organic matter decomposes to form humus. Plant and animal life builds the humus into the soil, aggregating it to form an open soil structure. The evolving soil infiltrates water rapidly, reinforcing the short ponding time that it was originally given.

In a basin with a long initial ponding time, the basin's soil becomes anoxic at the bottom of the pond. Peat forms from the incomplete decomposition of organic matter. The lack of soil oxygen inhibits plant rooting. The slimy peat seals soil voids, inhibiting further infiltration (Allison, 1947). The soil seals itself and further infiltration is prevented. The initially long ponding time becomes longer, and the cycle is repeated.

A soil evolves in the habitat that we create for it. Organic matter decomposition and plant rooting can work for infiltration, or against it, depending on the type of habitat that is created by initial ponding time. Where infiltration basins have failed with permanent ponding, the cause has often been biophysical soil development in saturated soils inside the basins. Residents and designers have attributed these failures to inflowing sediment, but in many cases incorrectly, because they did not know that the soil surface itself is dynamic.

Design to limit ponding time is the beginning of a basin's maintenance. In addition, the floor and sides must be densely vegetated to create a porous soil structure and self-maintaining soil ecosystem. Effective basin vegetation can be dense turf, dense herbaceous ground covers such as *Vinca* sp. or *Hedera helix*, or woody spreading ground covers such as *Juniperus* sp., *Leucothoe* sp., or *Euonymus* sp.

Where clogging has been caused not by soil development but by sediment from eroding soil in the drainage area, the cause can be easily traced, because the eroding soil and its muddy trail to the basin are visible. Removal of the foreign sediment layer from the basin may be called for, after the eroding sediment source has been stabilized.

Basin clogging can also be caused by surface crusting, where the soil surface is not protected by vegetative cover. Direct raindrop impact breaks up soil aggregates; dispersed mineral particles resettle into a uniform, sealing surface layer (Duley, 1939). This condition can be easily recognized by slicing through the soil surface with a penknife: the compact soil layer is only ¼ inch thick and is usually very distinct from the aggregated, porous soil below. On the surface, the soil between plants appears slick and shiny. This condition can be overcome by establishing a dense vegetative cover over the entire basin floor, using mulches or fabrics to prevent further raindrop impact while the cover is getting established.

INFILTRATION PONDING TIME

Because ponding time after a storm has a determining influence on the evolution of the soil surface, it is usually the single most important aspect of hydrologic design of infiltration basins.

The ponding time needed to maintain soil structure depends on the precipitation regime where a basin is located. Where rain is frequent year-round, as in the eastern part of the United States, short ponding times are needed to dispose of runoff and aerate the basin before the next rain occurs; currently a one-day (24-hour) limit is being used in Georgia. Where rain is infrequent year-round, as in the arid southwestern states, ponding time can be extended, perhaps up to seven days if the basin's vegetation can stand it, because there will still be time for the basin to aerate before another significant rain comes. Where rainfall is strongly seasonal, as in the Pacific states and southern Florida, intermediate ponding times can be used, because thorough aeration and biodegradation during the dry season may be sufficient to maintain infiltration for the rest of the year. Detailed observation of operating basins is necessary to refine ponding time limits in specific locales.

In basins that do not intercept the water table, ponding (drawdown) time after a storm is controlled by the soil's infiltration rate and the depth of ponding. With infiltration rate K in ft/day, basin depth D in ft, and ponding time t_p in days,

$$t_p = D \, / \, K$$

Where the soil has a given infiltration rate, the ponding depth that would produce a given ponding time can be derived by rewriting the preceding equation,

$$D = K \, t_p$$

According to this equation, infiltration basins with slowly permeable soil (low K) must be shallow in order to infiltrate the water within the given ponding time. Basins with highly permeable soil (high K) can be deep and still dispose the water they hold within the required time.

Infiltration rate K is also called saturated hydraulic conductivity (Ferguson, 1994). Its value depends on site-specific soil profiles. It can be estimated from soil borings or in situ infiltration rate tests. Infiltration tests involve expenses and do not always give consistent results. An alternative method is to identify the soil series from SCS soil surveys or on-site examination and then assign a value of K by association with the soil.

One indication of infiltration rate is soil texture, which is indicated in county surveys and can be confirmed by simple on-site investigation. Table 9.1 lists infiltration rates that can be assumed for soils that are unvegetated and have no aggregated structure.

Table 9.2 lists soil infiltration rates a different way. The soils are classified by SCS hydrologic soil group (HSG). Although there are only four soil groups, HSG is a relatively comprehensive way to classify soils, in the sense that it takes into account texture, structure, and all other conditions in the natural soil profile. The table also differentiates managed vegetation types, which give the soils different degrees of aggregation. Values for urban lawns may be similar to the table's pasture values. Values for woods may be higher than those shown for meadows. The table's range in values for each soil-vegetation combination reflects the range of site-specific conditions that can occur, such as extent of soil coverage

TABLE 9.1 Infiltration rates in unstructured soils, with soils classified by texture (Rawls et al., 1982)

Texture	Infiltration Rate (in./hr)	Infiltration Rate (ft/day)
Sand	8.27	16.54
Loamy sand	2.41	4.82
Sandy loam	1.02	2.04
Loam	0.52	1.04
Silt loam	0.27	0.54
Sandy clay loam	0.17	0.34
Clay loam	0.09	0.18
Silty clay loam	0.06	0.12
Sandy clay	0.05	0.10
Silty clay	0.04	0.08
Clay	0.02	0.04

by vegetation and degree of soil compaction by traffic. Take a close look at the figures in the table: in an unfavorable C soil, the infiltration rate with mature meadow vegetation is more than 15 times higher than that of an unstructured, unvegetated fallow soil!

DESIGN FOR THE SOIL PROFILE

Figure 9.7 shows examples of soil profiles. Like most soils, the examples have various horizons (layers) with different textures, structures, and infiltration rates. For the soil at your site, study the entire profile to identify both the horizons most limiting to infiltration and those most favorable for infiltration to take advantage of. Every SCS soil survey includes representative profiles of every soil series in the area, in the form of narrative descriptions; the examples in Figure 9.7 were drawn from such descriptions. On sites where soil borings have been taken, the borings can provide more conclusive site-specific information than do SCS's general surveys.

In specifying infiltration rate K to derive basin ponding time and depth, use K for the layer with the most limiting (lowest) value at or within a few feet below the floor of the

Table 9.2 Soil infiltration rates in ft/day, with soils classified by hydrologic soil group (derived from data in Musgrave and Holtan, 1964, pp. 12–26, and Nearing et al., 1996; data for D soils were not given by Nearing et al.).

	A	B	C
Fallow	0.44–0.66	0.22–0.44	0.07–0.22
Row crops (corn or beans in conventional tillage)	0.60–0.90	0.36–0.72	0.14–0.41
Wheat	0.81–1.21	0.47–0.94	0.18–0.54
Alfalfa	1.25–1.88	0.82–1.64	0.45–1.36
Pasture	1.60–2.40	0.95–1.90	0.44–1.31
Meadow	2.77–4.16	1.98–3.95	1.13–3.39

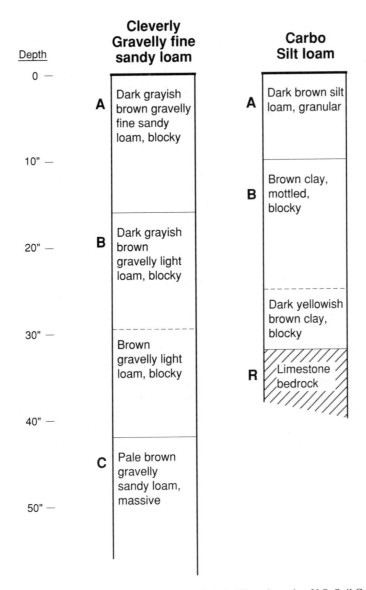

Figure 9.7   Soil profiles from the Salt Lake Valley of Utah (Cleverly series, U.S. Soil Conservation Service, 1972b, p. 74-75) and the Shenandoah valley of Virginia (Carbo series, U.S. Soil Conservation Service, 1982, pp. 87–88).

basin. Layers higher in the profile can be disregarded. Plan the excavation of a basin to take advantage of the most favorable horizons in the profile.

Shallow bedrock, groundwater, or impermeable soil layers may limit infiltration, no matter how high the infiltration rate of the overlying soil. Keep the floors of infiltration basins at least a few feet above any limiting layer (Figure 9.8). Where grading for roads or buildings has excavated below the original soil surface, take the finished grade elevation into account in setting the depths of basins in relation to subsurface layers. In some locales

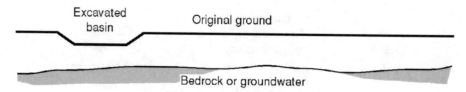

Figure 9.8 Potential subsurface conditions.

impermeable or slowly permeable soil layers overlie the permeable material into which you want to infiltrate; in these cases you can excavate the impermeable layer or provide a "drainage well" to pass water through it.

Where grading has filled above the original soil surface, avoid infiltrating into structural fills such as those that support roads or buildings. However, you can excavate a basin or well through a noncritical fill to infiltrate into the original soil.

ALTERNATIVE HYDROLOGIC OBJECTIVES

Early standards for infiltration set infiltration volume equal to the portion of design-storm runoff volume attributable to development. This is equal to the difference between Q_{vol}s before and after development (Juneja, 1974; Maryland Water Resources Administration, 1984). This standard is logically fair, because it requires developers to restore runoff volume by infiltrating exactly the amount of runoff attributable to the specific development. However, in terms of cost it is the worst of all possible situations: each development that complies with the standard exactly must have a system of adequate infiltration basins, at the same time that it must also have a second system of primary conveyances to discharge the portion of the runoff that is not infiltrated. This standard also eliminates the possibility of infiltration on many sites, because it is rigid: where the exact standard cannot feasibly be met, then the development does not attempt infiltration at all, but turns to other, less favorable methods of runoff treatment.

An alternative objective is to infiltrate the entire design-storm volume on sites where doing so is feasible, and on other sites to infiltrate the greatest amount possible within site-specific constraints. This flexible objective maximizes cost savings while infiltrating as much as possible on every site.

The most advantageous possible outcome of infiltration is to infiltrate the entire volume of design-storm runoff at the source. Where the entire volume is infiltrated, there is no overflow during the storm. The peak rate overflowing the basin is zero. This eliminates the need for primary conveyances downstream, as well as the cost that goes with them. Emergency (secondary) overflow is still needed as it is for every drainage system.

Where an infiltration basin infiltrates the entire flow volume,

$$A_b = Q_{vol}/D$$

where

$$A_b = \text{basin floor area, acres}$$

$$Q_{vol} = \text{storm runoff volume, af}$$

$$D = \text{basin depth, ft}$$

According to this equation, sites with slowly permeable soil, and therefore little depth, must be large in area in order to hold the total volume of runoff. Such basins consume valuable space on urban development sites. Consequently, the advantage of complete design-storm infiltration can be gained only on the most favorable sites, where highly permeable soils infiltrate large volumes of runoff with adequate speed.

On other sites, achieving the cost advantage of complete infiltration is less likely; primary swales or pipes are necessary to convey some of the runoff from the source during the design storm. This could occur where fine-textured soil limits the infiltration rate or where intense urban development leaves insufficient space for full-sized infiltration basins. Nevertheless, infiltration's environmental purposes of Q_{vol} reduction, water quality improvement, and restoration of subsurface flows and storages still deserve attention. And their attainment is still surprisingly feasible, because of the importance of small storms in the site's ecology.

Most of the rain that falls in an average year is in the small, frequent storms, not the big 10-year and 50-year design storms. Figure 9.9 shows the proportion of total annual precipitation that is infiltrated P_{inf}ann as a function of infiltration basin capacity in Atlanta and San Francisco. The curves are rather similar despite the difference in rainfall regimes between the two locations. The chart shows that a basin that infiltrates in one day the runoff from only 0.37 inches of rain (P) infiltrates half the rain that falls over the year. Basins that infiltrate the runoff from 1.8 inches of rain infiltrate 95 percent of the year's total. You do not have to infiltrate the entire volume of a 10-year or 50-year storm to restore most of a watershed's function. Despite the great size of a 10-year storm, it does little for groundwater in the long run, because it happens so infrequently. Most of the water that is

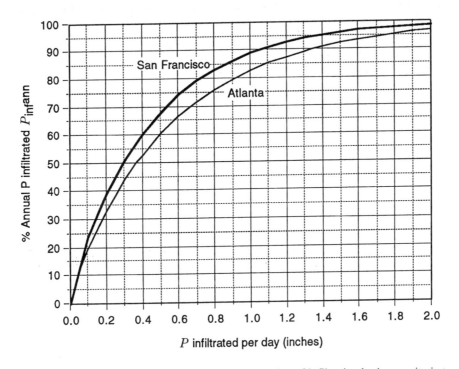

Figure 9.9 Portion of annual rainfall infiltrated as a function of infiltration basin capacity in two locations (based on daily rainfall 1950–1989).

available for groundwater and base flow replenishment is produced by small, frequent storms. The importance of small storms makes the infiltration of a large amount of cumulative rainfall highly feasible—on every site.

Restoration of water quality is equally independent of big design storms and peak flows. The "first flush" concentrates the pollution from pavements in the runoff from small, frequent storms and in the first small volume of runoff from large storms. Infiltration of small amounts of runoff is sufficient to solve the problem of pollution from impervious surfaces—on every site.

The big, flood-producing storms are destructive when they occur, but with years between their occurrence, they contribute little to total long-term rainfall, runoff, or water quality, and thus to a watershed's day-to-day ecological functions. Infiltrating only a small volume of water from each runoff event is sufficient to restore most of a watershed's environmental function. You should look for the chance to infiltrate at least some runoff on every development site, of every soil type and land use intensity.

INFILTRATION THROUGH PERMEABLE PAVEMENTS

Pavements are major surface covers in contemporary cities. Controlling pavement material to maintain infiltration is a basic step in the restoration of hydrologic function. Specific permeable materials for parking, traveling, and pedestrian surfaces were described in Chapter 2.

In porous material, water passes through the material's void spaces and is stored in the voids until it infiltrates the underlying soil. The quantity of void space controls the passage and storage of water. The quantity can be expressed as a proportion of the total pavement volume: with void volume and total volume in af, the void space V_d in af/af is given by

$$V_d = \text{void volume / total volume}$$

For a given required void volume to hold water, the total volume of the material is yielded by

$$\text{Total pavement volume} = \text{void volume} / V_d$$

For example, if a material's V_d is 0.4 af/af, then the total volume of the material must be 2.5 times the volume of water it is required to hold in its void space.

The same value of V_d relates equivalent depth of voids to the total depth of the material. With void depth and total depth in inches, the ratio V_d is given in in./in.:

$$V_d = \text{void depth / total pavement depth}$$

In a porous pavement of aggregate, asphalt, or concrete there is a base course of stone aggregate, with or without a binder. The base course is an infiltration basin where water is stored while infiltrating the underlying soil. The stone must be clean, open-graded crushed stone, such as No. 57 stone; this produces void space of 38 to 40 percent (0.38 to 0.40 af/af). A typical example of porous asphalt is the cross section in Figure 9.10; the principal features of porous concrete are similar.

The infiltrating soil surface is at the base of the pavement. Vegetation here is, of course, impossible, so the soil structure is free to break down without vegetation's ameliorating and

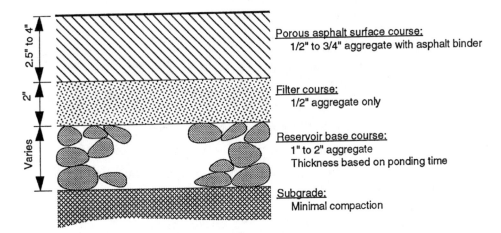

Figure 9.10   Standard porous asphalt pavement in the city of Rockville, Maryland.

rebuilding activity. It is prudent to assume a low rate of infiltration. To take this into account, after selecting a soil infiltration rate from Table 9.1 or 9.2, its value should be modified by a safety factor S_f between 0 and 1. The purpose of a safety factor is to take uncertainty into account. A value of 0.5 has been commonly used in practice (Ferguson, 1994). With this value of S_f, the infiltration rate effectively being used in design is half the rate that would be predicted from the physical characteristics of the soil alone.

In its simplest hydrologic function, porous pavement receives the rainfall that falls on it, without additional runoff from elsewhere on the site.

To infiltrate the entire design-storm rainfall, the depth of the pavement to hold the water is derived from the depth of rain P in inches and the void space V_d in in./in.:

$$\text{Pavement depth} = P \,/\, V_d$$

Whether this depth of rainfall can be infiltrated within a given ponding time t_p can be tested with the following equation, where K is in ft/day and t_p is in days:

$$t_p = P \,/\, (12\, K\, S_f)$$

If t_p found by the preceding equation is less than or equal to the acceptable ponding time, then the pavement design works: all design-storm P is infiltrated through the pavement rapidly enough, and no primary conveyance is needed downstream.

However, if t_p is greater than the acceptable ponding time, then the site's soil is not permitting the entire rainfall to be infiltrated fast enough. Some overflow must be permitted from the pavement and conveyed downstream. The pavement will operate like that shown in Figure 9.11. The pavement must be resized to hold only the amount that will infiltrate in the acceptable time.

The depth of water in inches that can be held for infiltrating within the acceptable ponding time can be found using

$$\text{Water depth to be held} = 12\, K\, S_f\, t_p$$

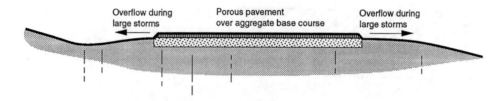

Figure 9.11   Operation of a porous pavement.

Because water occupies the void space in the pavement, the total depth of the base course where the water is stored is given by

$$\text{Pavement depth} = \text{water depth held} / V_d$$

The relative effect of this amount of infiltration on the design storm can be evaluated by finding the portion of rainfall held in the pavement:

$$\text{Portion of storm infiltrated} = \text{water depth held} / P$$

The amount of water overflowing the pavement and required to be conveyed downstream can be evaluated as a depth of water, in watershed inches:

$$\text{Amount of water overflowing} = P - \text{water depth held}$$

The effect of infiltrating a given amount of storm rainfall on the annual moisture flow can be evaluated with the chart in Figure 9.9. Enter the chart from the bottom with the depth of 24-hour rainfall held in the pavement and infiltrated within one day. Move vertically up to the line for the location with a rainfall regime closest to that at your site: San Francisco has 19.66 inches of precipitation per year, concentrated in a wet season; Atlanta has 48.71 inches per year, distributed essentially year-round. Interpolate between the two curves for intermediate types of rainfall regime. From the curve, move to the left to read the proportion of annual rainfall that the pavement infiltrates, P_{inf}ann.

Permeable Pavement Exercise

Exercise 9.1 finds the depth of pavement needed to infiltrate the rain that falls upon it. It compares the depths of pavement needed to infiltrate the entire design-storm precipitation and the amount that can infiltrate within an acceptable ponding time. The design storm is the same 24-hour rainfall used in Exercise 4.4.

Summary of Process

1. Obtain design-storm 24-hour rainfall P in inches, from Exercise 4.4.
2. Obtain maximum allowable ponding time t_p in days, from local standards.
3. Obtain soil infiltration rate K in ft/day from Table 9.1, Table 9.2, or on-site testing.
4. Select a value of infiltration rate safety factor S_f between 0.0 and 1.0.
5. Obtain base course void space V_d in in./in. (between 0.0 and 1.0) from specifications for the aggregate.

Exercise 9.1   Permeable pavement. The design storm is that same as that used in Exercise 4.4.

	Site 1	Site 2

Determining data

24 hour rainfall P
  (from Exercise 4.4)   = _____ in.   _____in.

Maximum allowable ponding
  time t_p (from local standards)   = _____ days   _____days

Soil infiltration rate K
  (from Table 9.1 or 9.2)   = _____ ft/day   _____ ft/day

Infiltration rate safety factor S_f
  (from local standards)   = _____   _____

Base course void space V_d
  (from material specs)   = _____ in./in.   _____ in./in.

Pavement infiltration of entire storm rainfall

Base course depth to hold P
  $= P / V_d$ = _____ in.   _____ in.

Ponding time t_p
  $= P / (12\,K\,S_f)$ = _____ days   _____ days

Portion of annual rainfall infiltrated
  (from Figure 9.9)   = _____ %   _____ %

Pavement infiltration limited by ponding time

Depth of water to be infiltrated
  $= 12\,K\,S_f\,t_p$ = _____ in.   _____ in.

Depth of base course to hold water
  $= $ depth of water $/ V_d$ = _____ in.   _____ in.

Portion of storm precipitation held
  $= $ depth of water $/ P$ = _____ in./in.   _____ in./in.

Depth of water overflowing pavement during
  storm $= P - $ depth of water   = _____ in.   _____ in.

Portion of annual rainfall infiltrated
  (from Figure 9.9)   = _____ %   _____ %

Pavement Infiltration of Entire Storm Rainfall

  6. Derive depth in inches of base course required to hold entire P, from pavement depth $= P / V_d$.

  7. Evaluate the result by finding the ponding time t_p in days, from $t_p = P / (12\,K\,S_f)$.

  8. Find the portion of annual rainfall infiltrated P_{inf}ann, in percentage, from Figure 9.9.

Pavement Infiltration Limited by Ponding Time

  9. Find the depth in inches of water that can be infiltrated within the acceptable ponding time, using $12\,K\,S_f\,t_p$.

  10. Find the depth in inches of base course to hold that amount of water, using depth of water held $/ V_d$.

  11. Evaluate the effect on the design storm by finding the portion of storm precipitation infiltrated in in./in., using depth of water held $/ P$. Also find the depth in inches of

precipitation that overflows the pavement and must be conveyed downstream, using P – depth of water held.

12. Evaluate the effect on the annual moisture flow by finding the portion of annual rainfall infiltrated P_{inf}ann, in percentage, from Figure 9.9.

Discussion of Results

1. What is the difference between the two sites in the proportion of storm precipitation that can be infiltrated? What differences in site conditions cause this difference in result?

2. Were you able to infiltrate the entire storm precipitation on either site? If so, what potential advantage does this give you in implementing the development project? In what specific ways would you take advantage of this benefit; how will your design look different as a result of it?

3. Did either site require overflow of some water during the design storm? Is infiltration through permeable pavement still a worthwhile thing to do? Why?

INFILTRATION BASINS

Figure 9.12 shows some of the essential features of an infiltration basin. The first flush of every storm enters the basin. Runoff from moderate and large storms that exceeds the capacity of the basin flows out the outlet. Water remaining in the basin is in contact with soil, and infiltrates. The duration of ponding is determined by the depth.

An open, vegetated basin like that shown in Figure 9.12 can be used where there is sufficient open space along the drainage system and occasional brief ponding is tolerable. It is immediately accessible for inspection and maintenance. Its construction involves little cost other than proper grading and planting. Vegetation actively maintains the soil's porous structure, so no safety factor on infiltration rate is required as it is for porous pavement.

A system of basins of this type can be found in the Maitland Center office development near Orlando, Florida. Grassed basins infiltrate each lot's runoff. The basins are integrated into each private lot's perimeter landscape of berms, plantings, and walkways. Curb cuts, sluices, and culverts deliver roof and parking lot runoff to the basins. The overflow from each basin drains into the next basin downstream. The overflow system can be traced to its end in a natural sinkhole lake at the site's lowest elevation.

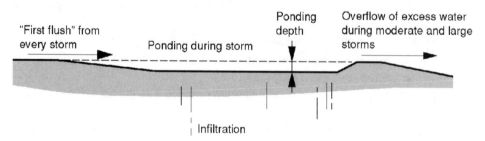

Figure 9.12 An infiltration basin.

Figure 9.13 Infiltration basin in the Buttonwood Cove residential development on Longboat Key, Florida.

Figure 9.13 shows a carefully maintained basin in the Buttonwood Cove residential development near Sarasota, Florida. The development left a low-lying grove of palm and cypress trees intact; only the ground cover was modified with grass and gravel. Runoff from roofs and pavements drains into the natural low area and infiltrates without disrupting the surface vegetation or the native trees. Because the area is oversized for the amount of runoff that reaches it, there is seldom a sign of standing water or wet soil.

A different type of basin is formed where it is excavated to intersect the water table, forming a permanent pool. Runoff entering the basin temporarily mounds on top of the water table until it infiltrates through the basin sides and by pushing down and spreading out the surrounding groundwater.

Excavated basins of this type ring a landfill in West Palm Beach, Florida, to treat potentially contaminated runoff and leachate before it enters adjacent native wetlands (Figure 9.14). To mitigate wetland destruction resulting from construction of the landfill, thousands of cypresses and other wetland plants were transplanted from the landfill construction area to the shores of the basins, where they comprise an artificial wetland accessible to wildlife and to the public.

Excavated basins at the office of Anaren Microwave near Syracuse, New York, are within a physiographically different context. The site is on a glacial lake plain with shallow groundwater and almost no topographic relief. Swampy parts of the site were excavated to generate fill for raising the building; the borrow areas became shallow ponds lined with native trees. The ponds' water levels fluctuate seasonally with the groundwater table. Runoff drains into the ponds, where it mounds up until it dissipates into the surrounding groundwater.

Figure 9.14   Excavated basin on the perimeter of the West Palm Beach, Florida, landfill.

The effect of an infiltration basin on annual moisture flow can be evaluated with the chart given in Figure 9.9, as can the effect of porous pavement. However, an infiltration basin receives runoff from elsewhere on the site, so before using the chart, you must convert the amount of infiltrating runoff into the precipitation that would produce it. For a given volume of water Q_{inf} in af that a basin can hold for infiltrating and its drainage area A_d in acres, convert the volume into equivalent depth of watershed runoff Q_d in inches, using $Q_d = 12\, Q_{inf} / A_d$. Use the SCS runoff chart in Figure 4.16 to find the storm rainfall that would produce that much runoff: enter the chart from the left side with the depth of runoff Q_d, move horizontally to the curve for your site's curve number, thence vertically downward to read the runoff-producing rainfall P. This is the "P infiltrated per day" in the scale at the bottom of the chart in Figure 9.9. Enter the chart from the bottom with the capacity of the infiltration basin in watershed inches; move up to the curve with the rainfall regime most similar to that at your site, thence to the left to read the proportion of annual rainfall infiltrated P_{inf}ann.

An infiltration basin can also be evaluated for its effect on peak rate of flow. An infiltration basin removes from the storm hydrograph a volume of water equal to the basin's capacity, beginning with the first runoff during the storm and continuing until the basin is full. The remaining runoff overflows at the rate it enters (unless there is a special additional provision for detention of this remaining volume). The chart in Figure 9.15 shows the effect on discharging q_p. The scale at the bottom is the proportion of the storm volume infiltrated, Q_{inf}/Q_{vol}. The scale at the left is the discharging peak flow as a proportion of the peak rate of the inflowing runoff, $q_p$out/$q_p$in. The curve shows that the effect on peak flow begins with infiltration of 37 percent of the flow volume. For basin capacities larger than that, the hydrograph is receding, and the effect of infiltration on q_p becomes greater with increasing Q_{inf}. When the entire flow volume has been captured ($Q_{inf}/Q_{vol} = 1$), the rate of overflow is zero.

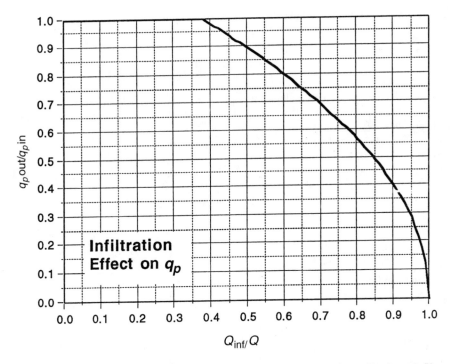

Figure 9.15 Peak rate overflowing an infiltration basin as a function of runoff volume infiltrated (based on data in U.S. Soil Conservation Service, 1972, Table 16.1).

Basin for Complete Infiltration Exercise

Exercise 9.2 sizes a vegetated basin to infiltrate the entire volume of design-storm direct runoff and evaluates its effects. The drainage area and design storm are the same as those used in Exercise 4.4. The site is also the same as that used in exercises for detention and extended detention, so the various approaches can be compared.

Summary of Process

1. From Exercise 4.4, obtain design-storm runoff volume Q_{vol} after development.
2. From local standards, obtain maximum allowable ponding time t_p in days.
3. From Table 9.1, Table 9.2, or on-site testing, obtain soil infiltration rate K in ft/day.

Basin Design

4. Derive maximum allowable basin depth D in ft, using $D = t_p K$.
5. Set infiltration volume Q_{inf} equal to Q_{vol}.
6. Find the basin's minimum floor area A_b in ac, using $A_b = Q_{inf} / D$.

Evaluation of Infiltration Effect

7. Obtain runoff-producing rainfall P in inches from the runoff estimate in Exercise 4.4.
8. Find portion of annual rainfall infiltrated P_{inf}ann from Figure 9.9.

Exercise 9.2 Vegetated basin to infiltrate entire runoff volume

	Site 1	Site 2
	Determining data	
Storm runoff volume Q_{vol} after development		
(from Exercise 4.4)	= _____ af	_____ af
Maximum allowable ponding		
time t_p (from local standards)	= _____ days	_____ days
Soil infiltration rate K		
(from Table 9.1 or 9.2)	= _____ ft/day	_____ ft/day
	Basin design	
Maximum basin depth D		
$= t K$	= _____ ft	_____ ft
Infiltration volume Q_{inf}		
$= Q_{vol}$	= _____ af	_____ af
Minimum basin floor area A_b		
$= Q_{inf} / D$	= _____ ac	_____ ac
	Evaluation of infiltration's effect	
Runoff-producing rainfall P		
(from Exercise 4.4)	= _____ in	_____ in
Portion of annual rainfall infiltrated		
(from Figure 9.9)	= _____ %	_____ %

Discussion of Results

1. What proportion of the storm flow volume Q_{vol} is infiltrated? What volume overflows the basin and must be conveyed downstream? What is the peak rate of discharge overflowing the basin?

2. What is the difference in basin area (A_b) between the two sites? What differences in site conditions cause the difference in basin size?

3. What portion of your site does the basin occupy (A_b / A_d)? In terms of land use allocation, is this so large as to be unreasonable? If so, you will be very interested in the next exercise, in which you design a basin that fits the available space on the site.

4. Is either basin so deep as to be unreasonable? If so, note the phrase "maximum basin depth" in the exercise. You are free to make the basin any depth less than the maximum. You can adapt the depth to the site and the land use in any way, as long as it does not exceed the maximum. As depth is reduced, what must happen to basin area A_b? Why?

Site-Limited Basin Exercise

Exercise 9.3 sizes a vegetated basin to infiltrate only the volume of runoff feasible within the constraints of the site's soil and land use and evaluates the effects. The drainage area and design storm are the same as those used in Exercise 4.4. The site is also the same as that used in Exercise 9.2, so the results of the two designs can be compared.

Exercise 9.3   Infiltration basin limited by site's space and soil.

	Site 1	Site 2

Determining data

Drainage area A_d
  (from Exercise 4.4)   = _____ ac   _____ ac
Curve number CN
  (from Exercise 4.4)   = _____   _____
Runoff volume Q_{vol}
  (from Exercise 4.4)   = _____ af   _____ af
Peak rate of runoff entering
  basin q_p (from Exercise 4.4)   = _____ cfs   _____ cfs
Area available for basin B
  (from site map)   = _____ ac/ac   _____ ac/ac
Maximum ponding time t_p
  (from local standards)   = _____ days   _____ days
Soil infiltration rate K
  (from Table 9.1 or 9.2)   = _____ ft/day   _____ ft/day

Basin design

Maximum basin depth D
  $= t_p K$ = _____ ft   _____ ft
Basin area A_b
  $= B A_d$ = _____ ac   _____ ac
Infiltration volume Q_{inf}
  $= D A_b$ = _____ af   _____ af

Evaluation of infiltration's effect

Volume of runoff overflowing basin
  during storm $= Q_{vol} - Q_{inf}$ = _____ af   _____ af
Portion of design storm runoff
  infiltrated $= Q_{inf} / Q_{vol}$ = _____ af/af   _____ af/af
Peak rate overflowing basin
  (from Figure 9.15)   = _____ cfs   _____ cfs
Volume infiltrated as depth of watershed
  runoff $D_{inf} = 12 Q_{inf} / A_d$ = _____ in   _____ in
Equivalent runoff-producing P
  (from Figure 4.16)   = _____ in   _____ in
Portion of annual rainfall infiltrated
  (from Figure 9.9)   = _____ %   _____ %

Summary of Process

1. From Exercise 4.4, obtain drainage area A_d in acres, curve number CN, runoff volume Q_{vol}, and peak rate of runoff q_p.
2. From a map of the site development, obtain proportion B of the drainage area available for a vegetated infiltration basin, in ac/ac.
3. From local standards, obtain maximum allowable ponding time t_p in days.

4. From Table 9.1, Table 9.2, or on-site testing, obtain soil infiltration rate K in ft/day.

Basin Design

5. Find maximum basin depth D in ft, using $D = t_p K$.
6. Find area available for basin A_b in ac, using $A_b = B A_d$.
7. Find the infiltration volume Q_{inf} in af that the basin can hold, using $Q_{inf} = D A_b$.

Evaluation of Infiltration's Effect

8. Find the volume of runoff in af overflowing the basin during a storm, using $Q_{vol} - Q_{inf}$.
9. Find the portion of the design storm infiltrated in af/af, using Q_{inf} / Q_{vol}.
10. Find the peak rate overflowing the basin from Figure 9.15.
11. Convert volume infiltrated to equivalent depth of watershed runoff Q_d in inches, using $Q_d = 12 Q_{inf} / A_d$, and use the result to find equivalent runoff-producing precipitation P in inches from Figure 4.16. Use this amount to find portion of annual rainfall infiltrated P_{inf}ann from Figure 9.9.

Discussion of Results

1. Which site requires the larger basin? What site and climatic conditions contribute to the requirement of such a large basin?
2. Compare the peak rate of flow discharging from your basin with the peak rate before development, which you found in Exercise 4.3. In what ways has infiltration contributed to reduction in downstream peak flows? In what ways has it failed to contribute to reduction? Do you think that detention is necessary on this site, in addition to infiltration? Why?
3. Compare the runoff volume discharging from your basin with the runoff volume before development, which you found in Exercise 4.3. In your judgment, has infiltration adequately reduced runoff volume? Why? In what ways could you further reduce runoff volume on this site?
4. Imagine that a contractor is installing your basin. What features of the installed basin and its surroundings should you check, because they are essential to the basin's ability to operate correctly? Describe how would you do that, working alone and using only equipment that you can carry in your pocket or in your hand.
5. Describe what would happen if your basin receives sediment from eroding soil in its drainage area, and what you would do about it.

SUBSURFACE BASINS

Where space is limited because of the intensity of development on a given site, basins can be constructed under the land surface, leaving the surface for functional use such as parking.

The storage capacity of underground basins is created with open-graded aggregate, as it is under porous pavements (Figure 9.16). Capacity is sometimes supplemented with perforated pipes or premanufactured chambers. Wrapping deep vertical sides in filter fabric prevents surrounding soil from collapsing into stone voids. The cost of subsurface construction is substantial as compared with that of open vegetated basins, but in intense land uses the high cost of land and the demand for functional surface use make the additional construction cost a feasible investment.

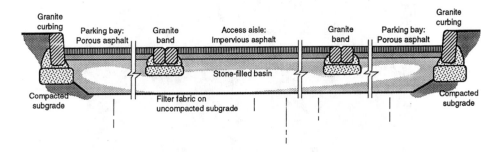

Figure 9.16 Subsurface basin at Morris Arboretum, Philadelphia (after Strom and Nathan, 1993, p. 202).

Use drainage inlets to let runoff into the basin, where it does not enter through the aggregate surface or through porous pavement. The inlets' perforated walls or attached perforated pipes admit water into the stone voids. Removable inlet grates allow access for inspection and maintenance.

In Glen Burnie, Maryland, the parking lot at the Cromwell Field shopping center is built over a stone-filled basin six feet deep and an acre in area. All that is visible from the surface is conventional drainage inlets. Despite the coverage of more than 90 percent of the site with impervious surfaces, the hydrologic function of the site has been restored as if the development had never happened.

Near Orlando, Florida, the tennis courts at the Sunbay Club multifamily residential development are underlain by a gallery of perforated pipes (Figure 9.17). Runoff enters the pipes from the surrounding stormwater collection system. The pipes are surrounded by

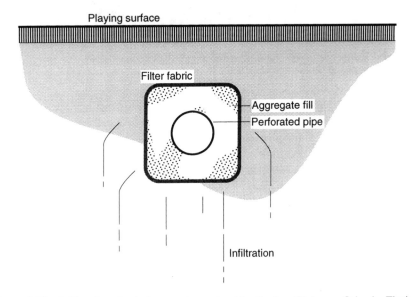

Figure 9.17 Infiltration pipe below tennis courts at the Sunbay Club near Orlando, Florida.

Figure 9.18   Infiltration basin and "well" at a church near Spokane, Washington.

stone to enlarge the storage volume and infiltration area. Infiltrating water recharges the Floridan aquifer, which is the principal water supply for people in the region. Before reaching the infiltration gallery, runoff passes through a series of ornamental fountains and pools that help aerate the water and remove sediment.

In Spokane, Washington, subsurface "wells" supplement the capacities of grassed infiltration basins. A typical combination is shown in Figure 9.18. The basins are sized to capture and infiltrate the first half-inch of runoff, filtering it through the vegetated soil. When large storms occur, the overflow goes over the rim of the stone-filled well and directly recharges the underlying gravel aquifer. Spokane's basins and wells are designed and monitored with great care to recharge large quantities of runoff without degrading water quality, because the underlying aquifer is the sole source of water for people in the urban area.

The hydrologic design of a subsurface basin combines elements that you have already experienced in design of porous pavements (Exercise 9.1) and infiltration basins (Exercises

9.2 and 9.3). Runoff enters the basin from the drainage area. The aggregate's void space V_d where infiltrating water is stored must be taken into account in finding the basin's volume. Vegetation is impossible on the infiltrating soil surface, so a safety factor S_f on infiltration rate is prudent. The infiltrated volume can be either the total volume of storm runoff or the amount limited by the space for a basin on the site.

Subsurface Basin Exercise

Exercise 9.4 sizes a subsurface basin to infiltrate the volume of storm runoff feasible within the constraints of the site's size and soil and evaluates the effects. The basin is constructed of open-graded stone aggregate. The drainage area and design storm are the same as those used in Exercise 4.4. The site is also the same as that used in Exercise 9.3, so the results of vegetated and subsurface basins can be compared.

Summary of Process

1. From Exercise 4.4, obtain drainage area A_d in acres, curve number CN, runoff volume Q_{vol}, and peak rate of runoff q_p.
2. From a map of the site development, obtain proportion B of the drainage area available for a subsurface basin, in ac/ac.
3. From local standards, obtain maximum allowable ponding time t_p in days and infiltration rate safety factor S_f.
4. From Table 9.1, Table 9.2, or on-site testing, obtain soil infiltration rate K in ft/day.
5. From construction specifications, obtain stone void space V_d in af/af.

Basin Design

6. Find maximum allowable depth in ft of water to be held in basin, using $T_p K S_f$.
7. Find total depth of basin in ft, from depth of water held / V_d.
8. Find area available for basin A_b in ac, using $A_b = B A_d$.
9. Find infiltration volume Q_{inf} in af that the basin can hold, using Q_{inf} = depth of water held $\times A_b$.
10. Find total basin volume including both water and stone, in af, using depth of basin $\times A_b$.

Evaluation of Infiltration's Effect

11. Find volume of runoff in af overflowing basin during storm, using $Q_{vol} - Q_{inf}$.
12. Find portion of design storm infiltrated in af/af, from Q_{inf} / Q_{vol}.
13. Find peak rate overflowing basin from Figure 9.15.
14. Convert volume infiltrated to equivalent depth of watershed runoff Q_d in inches, from $Q_d = 12 Q_{inf} / A_d$, and use the result to find equivalent runoff-producing precipitation P in inches from Figure 4.16. Use this amount to find portion of annual rainfall infiltrated P_{inf}ann from Figure 9.9.

Discussion of Results

1. Is either basin unrealistically deep? If so, note the phrase "maximum depth." You are free to set the depth at any value less than the maximum. For the general topography, soil, and land use of your site, what approximate depth do you think would be realistic and appropriate, without exceeding the maximum?

Exercise 9.4   Subsurface infiltration basin

	Site 1	Site 2

Determining data

Drainage area A_d
  (from Exercise 4.4)   = _____ ac   _____ ac
Curve number CN
  (from Exercise 4.4)   = _____   _____
Runoff volume Q_{vol}
  (from Exercise 4.4)   = _____ af   _____ af
Peak rate of runoff entering basin
  q_p (from Exercise 4.4)   = _____ cfs   _____ cfs
Area available for basin B
  (from site map)   = _____ ac/ac   _____ ac/ac
Maximum allowable ponding time
  t_p (from local standards)   = _____ days   _____ days
Infiltration rate safety factor S_f
  (from local standards)   = _____   _____
Soil infiltration rate K
  (from Table 9.1 or 9.2)   = _____ ft/day   _____ ft/day
Stone void space V_d
  (from material specs.)   = _____ af/af   _____ af/af

Basin design

Maximum allowable depth of water
  in basin = $t_p\, K\, S_f$ = _____ ft   _____ ft
Maximum total depth of basin
  = depth of water / V_d = _____ ft   _____ ft
Basin area A_b
  = $B\, A_d$ = _____ ac   _____ ac
Infiltration volume Q_{inf}
  = depth of water $\times A_b$ = _____ af   _____ af
Total basin volume
  = depth of basin $\times A_b$ = _____ af   _____ af

Evaluation of infiltration's effect

Volume of runoff overflowing basin
  during storm = $Q_{vol} - Q_{inf}$ = _____ af   _____ af
Portion of design storm runoff
  infiltrated = Q_{inf} / Q_{vol} = _____ af/af   _____ af/af
Peak rate overflowing basin
  (from Figure 9.15)   = _____ cfs   _____ cfs
Volume infiltrated as depth of watershed
  runoff $Q_d = 12\, Q_{inf} / A_d$ = _____ in.   _____ in.
Equivalent runoff-producing P
  (from Figure 4.16)   = _____ in.   _____ in.
Portion of annual rainfall infiltrated
  (from Figure 9.9)   = _____ %   _____ %

2. Assume that the surface over your basin is used for a double-loaded, 90° parking lot. What are the typical dimensions of a parking space in your locale, and how wide is the traveling lane between spaces? How many square feet of parking lot does each parking space occupy, counting both the stall and the adjacent traveling lane? Calculate the number of cars that could park on the surface over your basin. Identify a parking lot near you that might hold a similar number of cars. Go and look at it to get a feel for the dimensions and physical character of the land use you are planning. Comment on its needs for landscape design.

3. In your locale, what is a typical number of parking spaces to allocate per dwelling unit in a multifamily development? Given the number of cars that could park on the surface of your basin, how many dwelling units are being supported? If these are all the dwelling units in your site, what is the density of dwelling units per acre? Is this a reasonable density in terms of economic investment, land use intensity, and open space preservation?

4. If your basin did not hold the entire volume of runoff during the design storm, is infiltration still worth using on this site? What are its advantages and disadvantages? If you do not use infiltration at all, should you use any other kind of runoff control instead?

WATER BALANCE EVALUATION

The quantity of precipitation falling on a site in an average year is a resource that is apportioned to various components of the environment. The long-term partitioning of flow between direct runoff, evapotranspiration, and base flow is of great concern to groundwater recharge, flooding, recreation, aquatic life, and water supplies.

Previous exercises have evaluated infiltration's long-term effects by finding the proportion of annual rainfall infiltrated. A more complete image of the long-term effect can be produced by including infiltration in a calculation of the Thornthwaite water balance, making a complete accounting of hydrologic effects on the landscape and their variations with the seasons. Creating direct runoff with impervious surfaces, then infiltrating some of it in basins, could have unexpected seasonal effects on a site's environment through evapotranspiration, soil moisture, groundwater, and base flow. A complete hydrologic restoration would be found where infiltration causes base flow to equal that before development. In other words, there would be as much water returned to the site after development as was there before development.

The evaluation can be done by applying the long-term infiltration ratio P_{inf}ann, determined from Figure 9.9, to the monthly amounts of direct runoff in a water balance calculation. An infiltration basin captures the direct runoff that has been generated. Based on the infiltration ratio, the basin redirects some of the runoff to the soil and leaves the overflow as direct runoff. The rest of the water balance is completed normally, with these slightly redirected throughflows.

Water Balance Evaluation Exercise

Exercise 9.5 uses the monthly water balance to evaluate the effect of the vegetated basin that was designed in Exercise 9.3. The monthly precipitation and direct runoff after development are those recorded in Exercise 5.2.

Exercise 9.5 Stormwater infiltration and remaining runoff. All monthly values are in inches. Complete this exercise for both Site 1 and Site 2.

	Jan.	Feb.	Mar.	Apr.	May	June	July	Aug.	Sept.	Oct.	Nov.	Dec.

Site data

Proportion of annual precipitation
infiltrated P_{inf}ann (from Exercise 9.3) = _____ in./in.

Infiltration and remaining direct runoff

Direct runoff Q_d after development
(from Exercise 5.2) = ___ — — — — — —

Infiltration of direct runoff
= $Q_d \times P_{\text{inf}}$ann = ___ — — — — — —

Remaining direct runoff
= $Q_d \times (1 - P_{\text{inf}}\text{ann})$ = ___ — — — — — —

Summary of Process

1. Obtain from Exercise 9.3 the proportion of annual precipitation infiltrated P_{inf}ann in percentage. Convert the amount to in./in. by dividing by 100.
2. For each month, obtain from Exercise 5.2 the monthly direct runoff Q_d after development, in in./mo.
3. Find the monthly direct runoff infiltrated, using $Q_d \times P_{inf}$ann.
4. Find the monthly direct runoff remaining after infiltration, using $Q_d \times (1 - P_{inf}$ann$)$.
5. Recalculate Exercise 5.4, using the remaining direct runoff that you just found in place of the total direct runoff Q_d. Complete the water balance by recalculating Exercise 5.5.

Discussion of Results

1. Compare the results of Exercise 9.5 with the after-development results found in the original calculation of Exercises 5.4 and 5.5. Specifically compare the site's outflows of evapotranspiration, direct runoff and base flow, and its storages in soil moisture and groundwater. To what degree has infiltration altered the hydrology of the previously unmitigated development site? Considering the ecological roles of these flows and storages, is this amount of infiltration a step in the right direction? Is it worth doing?
2. Compare the results with the before-development results found in Exercises 5.4 and 5.5. If the ultimate objective of infiltration is to recharge water into subsurface soil moisture and groundwater, do you think it has accomplished this to a satisfactory degree? How could the basin or the site development be revised to increase the proportion that infiltrates?

SUMMARY AND COMMENTARY

Infiltration, like conveyance and detention, is a form of runoff disposal because it discharges stormwater to the environment. However, it is qualitatively different because it discharges moisture to its natural flow path in the subsurface. Given that excess runoff has been generated on an urban site, conveyance and detention manage stormwater; infiltration restores it. Infiltrated water never aggravates downstream floods. It supports groundwater and base flows. Soil filtration protects the quality of aquifers and streams.

Restorative infiltration starts at the potential sources of runoff, with porous pavements. It continues wherever possible with vegetated swales and basins that bring runoff continuously and repeatedly into contact with vegetation and soil.

Porous pavements and infiltration basins are facilities in a site development program. Appropriate spaces must be reserved for them in site plans. Their dimensions must be integrated with the development's functional layout and grading plans.

REFERENCES

Allison, L. E., 1947, Effect of Microorganisms on Permeability of Soil under Prolonged Submergence, *Soil Science* vol. 63, pp. 439–450.

Aronson, D. A. and G. E. Seaburn, 1974, *Appraisal of Operating Efficiency of Recharge Basins on Long Island, New York*, Water-Supply Paper 2001-D, Washington: U.S. Geological Survey.

Duley, F. L., 1939, Surface Factors Affecting the Rate of Intake of Water by Soils, *Soil Science Society of America Proceedings* vol. 4, pp. 60–64.

Engstrand, Daniel, 1983, *Retention Ponds: Analysis and Design of Ponds Without Outlets*, St. Paul, Minn.: Minnesota Department of Transportation.

Ferguson, Bruce K., 1990, Role of the Long-term Water Balance in Management of Stormwater Infiltration, *Journal of Environmental Management* vol. 30, pp. 221–233.

Ferguson, Bruce K., 1994, *Stormwater Infiltration*, Boca Raton: Lewis Publishers.

Ferguson, Bruce K., 1995, Storm-Water Infiltration for Peak-Flow Control, *Journal of Irrigation and Drainage Engineering* vol. 121, no. 6, pp. 463ñ466.

Ferguson, Bruce K., 1996, Preventing the Problems of Urban Runoff, *Renewable Resources Journal* vol. 13, no. 4, pp. 14–18.

Florida Concrete and Products Association, (n. d.), *Pervious Pavement Manual*, Orlando: Florida Concrete and Products Association.

Hannon, Joseph B., 1980, *Underground Disposal of Storm Water Runoff*, Design Guidelines Manual, FHWA-TS-80-218, Washington: U.S. Federal Highway Administration.

Juneja, Narendra, 1974, *Medford: Performance Requirements for the Maintenance of Social Values Represented by the Natural Environment of Medford Township, N.J.*, Philadelphia: University of Pennsylvania Department of Landscape Architecture and Regional Planning, Center for Ecological Research in Planning and Design.

Ku, Henry F. H. and Dale L. Simmons, 1986, *Effect of Urban Stormwater Runoff on Groundwater Beneath Recharge Basins on Long Island, New York*, Water-Resources Investigations Report 85-4088, Syosset, N. Y.: U.S. Geological Survey.

Ku, Henry F. H., Nathan W. Hegelin, and Herbert T. Buxton, 1992, Effects of Storm-Runoff Control on Ground-Water Recharge in Nassau County, New York, *Ground Water* vol. 30, no. 4, pp. 507–514.

Lindsey, Greg, Les Roberts, and William Page, 1992a, Maintenance in Stormwater BMPs in Four Maryland Counties: A Status Report, *Journal of Soil and Water Conservation* vol. 47, no. 5, pp. 417–422.

Lindsey, Greg, Les Roberts, and William Page, 1992b, Inspection and Maintenance of Infiltration Facilities, *Journal of Soil and Water Conservation* vol. 47, no. 6, pp. 481–486.

Maryland Water Resources Administration, 1984, *Standards and Specifications for Infiltration Practices*, Annapolis: Maryland Department of Natural Resources.

Musgrave, G. W., and H. N. Holtan, 1964, Infiltration, Section 12 in *Handbook of Applied Hydrology*, Ven Te Chow, editor, New York: McGraw-Hill.

Nearing, M. A., B. Y. Liur, L. M. Risse, and X. Zhang, 1996, Curve Numbers and Green-Ampt Effective Hydraulic Conductivities, *Water Resources Bulletin* vol. 32, no. 1, pp. 125–136.

Nightingale, Harry I., 1987a, Accumulation of As, Ni, Cu, and Pb in Retention and Recharge Basins Soils from Urban Runoff, *Water Resources Bulletin* vol. 23, no. 4, pp. 663–672.

Nightingale, Harry I., 1987b, Water Quality Beneath Urban Runoff Management Basins, *Water Resources Bulletin* vol. 23, no. 2, pp. 197–205

Nightingale, Harry I., and W. C. Bianchi, 1973, Groundwater Recharge for Urban Use, *Ground Water* vol. 11, No. 6, pp. 36–43.

Nightingale, Harry I., and W. C. Bianchi, 1977, *Environmental Aspects of Water Spreading for Ground-water Recharge*, Technical Bulletin No. 1568, Washington: U.S. Department of Agriculture, Science and Education Administration.

Pitt, Robert, Shirley Clark, Keith Parmer, and Richard Field, 1996, *Groundwater Contamination from Stormwater Infiltration*, Chelsea, Mich.: Ann Arbor Press.

Rawls, W. J., D. L. Brakensiek, and K. E. Saxton, 1982, Estimation of Soil Water Properties, *Transactions of the American Society of Agricultural Engineers*, vol. 25, no. 5, pp. 1316–1320, 1328.

Seaburn, G. E,. and D. A. Aronson, 1974, *Influence of Recharge Basins on the Hydrology of Nassau and Suffolk Counties, Long Island, N.Y.*, Water-Supply Paper 2031, Washington: U.S. Geological Survey.

Smith, T. W., R. R. Peter, R. E. Smith, and E. C. Shirley, 1969, *Infiltration Drainage of Highway Surface Water,* Research Report 6328201, Sacramento: California Department of Transportation.

Sorvig, Kim, 1993, Porous Paving, *Landscape Architecture* vol. 83, no. 2, pp. 66–69.

Strom, Steven, and Kurt Nathan, 1993, *Site Engineering for Landscape Architects*, second edition, New York: Van Nostrand Reinhold.

Thelen, Edmund, and L. Fielding Howe, 1978, *Porous Pavement*, Philadelphia: Franklin Institute Press.

U.S. Soil Conservation Service, 1972a, *National Engineering Handbook,* Section 4, Hydrology, SCS/ENG/NEH-4, Washington: U.S. Soil Conservation Service.

U.S. Soil Conservation Service, 1972b, *Soil Survey of Utah County, Utah, Central Part*, Washington: U.S. Soil Conservation Service

U.S. Soil Conservation Service, 1982, *Soil Survey of Rockingham County, Virginia.* Washington: U.S. Soil Conservation Service

CHAPTER 10

WATER HARVESTING

Water harvesting is the collection of runoff for direct use. The term *water harvesting* originated in the use of runoff for irrigation in the arid Southwest; the same term can be used for the supply of water to ponds or any other uses. Water harvesting can be supplemented by management of excess runoff leaving a site, where it is explicitly designed for that.

Water harvesting requires a change of viewpoint. Instead of assessing drainage areas as sources of nuisance runoff, water harvesting plans catchment areas to generate water supplies. Instead of minimizing runoff impact, it utilizes runoff efficiency. Instead of diverting runoff away from the places where it could do damage, it directs it to points of use.

The drainage area, or catchment area, is the source of water supply, and the harvest area is the point of use (Figure 10.1). Monthly flows through the catchment and harvest areas are estimated by the water balance, as described in Chapter 5.

Inflow to the harvest area is a monthly volume of water discharging from the catchment area. It is equal to Q_d, direct runoff, where the catchment area is small and above the springs at the sources of seasonal and perennial flows. It is equal to $q\Sigma$, the sum of direct runoff and base flow, where water is harvested from streams with seasonal or perennial flow between storms.

In most landscape applications, the harvest area's outflows are evapotranspiration and infiltration. Their rates are measured as depth (inches) per month. Their volumes increase with the size of the harvest area, because volume = area × depth (Figure 10.2).

The balance between inflow from the catchment area and outflows (demands or losses) during use limits the sizes of ponds and irrigated areas that can be supported. With inflow in ac-in./mo and outflows in in./mo, the supportable harvest area in acres is the area that equalizes outflow with inflow:

Supportable harvest area = inflow volume / outflow depth

223

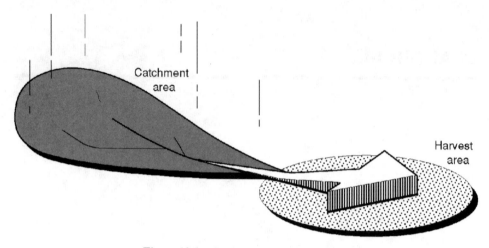

Figure 10.1   Catchment and harvest areas.

Direct precipitation onto a harvest area supplements inflow from the catchment area. However, like the outflows, it is measured in depth rather than volume; the volume of water it contributes grows with the size of the harvest area. Consequently, direct precipitation is counted as a reduction of losses rather than an addition to outflow, by subtracting it from actual losses such as evaporation and infiltration.

The quality of harvested water depends on the land uses in the catchment area. Dumpster pads, service areas where trash is handled, and heavily used parts of roads and parking lots can be intense generators of pollutants. Most roofs, turf that is not excessively maintained with chemicals, and pavements with only light vehicular traffic generate relatively clean runoff that can be suitable for pond or irrigation water supplies. The precise quality of roof runoff depends on the roofing material: wooden shingles treated with preservatives release chemical contaminants; asphalt shingles produce somewhat cleaner runoff; the best quality comes from tile and terra-cotta roofs (Arizona Water Resource, 1994). A site can be laid out and graded to direct runoff to a harvest area selectively from clean sources. Unavoidable runoff of questionable quality should be run through a biofilter swale before delivery to a sensitive application.

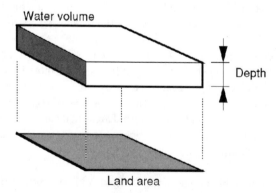

Figure 10.2   Relationship of water depth and volume to the land area where water is applied.

SUPPORT OF PERMANENT POOLS

Permanent pools are the essential components of ponds or wetlands for wildlife, aesthetics, recreation, water quality enhancement, or cooling water. The water supply to a pool maintains the water level against losses to evaporation and infiltration.

The rate of evaporation from a pool can be assumed equal to potential evapotranspiration Et_o. Monthly values of Et_o can be estimated by the Thornthwaite method, as described in Chapter 5, or taken from local evapotranspiration monitoring where it exists.

The soil infiltration rate K can be estimated from Table 9.1, which lists unstructured soils such as a those that might occur at the bottom of a pool. To assure a permanently wet regime amid the uncertainty surrounding natural infiltration rates, the value of K should be increased by a safety factor S_f. A common value of S_f in practice is 2.0; in other words, the infiltration rate assumed in water harvesting design is twice that indicated by the physical properties of the soil alone.

The following equation relates the specific flows to estimate the maximum supportable pool area in any one month:

$$\text{Maximum supportable } A_p = \text{inflow} / (Et_o + K\,S_f - P)$$

where

$$A_p = \text{pool area, ac}$$

$$\text{inflow} = \text{direct runoff } Q_d \text{ or total flow } q\Sigma, \text{ ac-in./mo}$$

$$Et_o = \text{potential evapotranspiration, in./mo}$$

$$K = \text{soil infiltration rate, in./mo}$$

$$S_f = \text{safety factor on infiltration rate, no units}$$

$$P = \text{precipitation, in./mo}$$

Apply the preceding equation to each of the 12 months to find the smallest area supportable at any time of the year. The smallest of the monthly pool areas is the maximum that is supportable year-round. During the other 11 months, a water surplus exists and excess inflows are lost through the pond's outlet.

If a pond is deliberately sealed with impervious material such as plastic sheeting or a bentonite clay blanket, the infiltration rate approaches zero and can be deleted from the equation.

The preceding equation can yield a negative pool area in months when P is very large. This means that in those months, it is impossible to dispose all inflow without spilling out excess from the pool's outlet. In months when this occurs, set the supportable pool area infinitely large.

If the pool is doubling for flood control, then the flood detention storage is located above the pool's surface. The rate of outflow during storms is determined by the outlet's design and by stage above the pool. The volume available for flood storage can be found by applying the contour planes method to a grading plan. The lowest "contour" in the flood storage volume is the edge of the permanent pool; its area is A_p. Given the volume and stage, an outlet can be selected that releases the required outflow rate, as described in Chapter 7.

Near Columbus, Ohio, at the headquarters of the Chemlawn corporation, runoff from roofs and lawns is supplied to a delicately detailed and lavishly planted pond located prominently near the building. James Bassett's planning of the site identified topographic drainage divides. Parking lots and roadways that might produce pollutants were located outside the pond's catchment area. The catchment was contoured to further segregate it from roads while delivering an adequate supply of runoff to the pond. The outlet from the pond discharges freely into a lower pond, farther from the building, where large storm flows are detained.

On the campus of Hofstra University on Long Island, the floor of a "recharge" (infiltration) basin was lined with a plastic sheet and planted with native wetland plants (Figure 10.3). The lined area holds a permanent pool between storm events. Runoff from an adja-

Figure 10.3   The wildlife sanctuary on the campus of Hofstra University, Hempstead, New York, consisting of an infiltration basin that was lined with plastic sheeting to contain water between storm events.

cent highway is filtered to a degree by a biofilter swale before reaching the pool. When runoff enters the pool, the water level rises and excess water infiltrates through the basin's unlined sides. Biological surveys have documented extraordinarily diverse bird life, as compared with the surrounding highly urbanized area. The university maintains the site as part of its arboretum and has designated it a wildlife sanctuary.

In Waterloo, Iowa, the 15-acre site of Northwest Mortgage supplies runoff to a one-acre permanent pool prominently located between the wings of the building. The pool provides the building's cooling water through an array of recirculating spray fountains. An edging of gravel and placed boulders accommodates up to two feet of water-level fluctuation for flood storage. The fountainheads are on risers to allow the cooling system to function even when the storm storage is highest. The reservoir reduces the design storm's 62-cfs inflow to 5 cfs so as not to overload a preexisting 12-inch city storm sewer. A well resupplies the pond in some summer months, when rapid evaporation is aggravated by a high cooling load on the spray fountains.

Analytical Exercise

1. Using algebra, derive an equation for the amount of supplemental well or municipal water necessary to supply a pond of a given size when the natural supply from the catchment area is not sufficient. Begin by adding a term for the supplemental supply to the equation for supportable area. Then solve the equation for that term.

Permanent Pool Exercise

Exercise 10.1 checks to see whether the permanent pool required for extended detention in Exercise 8.2 can be supported year-round by the inflow from the catchment area. The site is also the same as that used for water balance exercises in Chapter 5.

Summary of Process

1. From Exercise 8.2, obtain drainage area A_d in acres.
2. From Table 9.1, obtain soil infiltration rate K in ft/day. If the bottom of the pool is sealed, the value of K is zero.
3. From local standards, obtain infiltration safety factor S_f, equal to or greater than 1. Complete the remaining steps for each month separately.

Pool Inflow

4. From Exercise 5.5, obtain inflow Q_d or $q\Sigma$ in in./mo. Convert the amount into ac-in./mo from $Q_d A_d$ or $q\Sigma A_d$.

Pool Outflow

5. From Exercise 5.3, obtain potential evapotranspiration Et_o in in./mo.
6. Find pool infiltration K_p in in/mo from $360 K S_f$.
7. From Exercise 5.1 or 5.2 obtain direct precipitation P (not P_{net} after snow storage) in in/mo.
8. Sum the effective outflows, $Et_o + K_p - P$.

Supportable Pool Area

9. Find supportable pool area A_p in ac, using A_p = inflow / sum of outflows.

Exercise 10.1 Supportable permanent pool area. Complete this exercise for both Site 1 and Site 2.

	Jan.	Feb.	Mar.	Apr.	May	June	July	Aug.	Sept.	Oct.	Nov.	Dec.
					Site data							
Drainage area A_d (from Exercise 8.2)	= _____ ac											
Soil infiltration rate K (from Table 9.1)	= _____ ft/day											
Infiltration safety factor S_f (from local standards)	= _____											
					Pool inflow							
Monthly inflow Q_d or $q\Sigma$, in./mo (from Exercise 5.5)	=					___				___		
Monthly inflow volume, ac-in./mo $= Q_d A_d$ or $q\Sigma A_d$	=					___				___		
					Pool outflows							
Potential evapotranspiration Et_o, in./mo (from Exercise 5.3)	=			___		___				___		
Pool infiltration K_p, in./mo $= 360\, K\, S_f$	=			___		___				___		
Direct precipitation P, in./mo (from Exercise 5.1 or 5.2)	=			___		___				___		
Total outflows, in./mo $= Et_o + K_p - P$	=			___		___				___		
					Supportable pool area							
Monthly supportable pool area A_p, ac $=$ inflow / total outflows	=			___		___				___		
Year-round supportable pool area $=$ smallest monthly A_p	= _____ ac											

228

10. Identify the smallest of the monthly supportable pool areas as the maximum area that is supportable year-round.

Discussion of Results

1. Compare the supportable pool area to the area of permanent pool required for extended detention, as found in Exercise 8.2. Is a pool of the required size supportable year-round on your site?

2. If a pool of the required size is not supportable, but you build a pond of that size anyway, in what ways will the pond's water level and throughflow fluctuate with the seasons?

3. If a pool of the required size is not supportable, and you build a pond of only the supportable size, what will be the effect on residence time and sediment trap efficiency?

4. If a pool larger than the required size is supportable, and you build a pond of the full supportable size, what will be the effect on residence time and sediment trap efficiency? What will be the effect on dissolved constituents?

5. For each site, estimate the area of the supportable pool as a proportion of its catchment area (A_p/A_d). Which site is able to support the larger pool? What site and climatic conditions contribute to the support of such a large pool?

6. In your judgment, is the supportable pool area on either site so small as to be unreasonable in terms of land use and landscape design? What approaches in site design could you take to overcome this difficulty?

7. What happens to the excess inflow in the 11 months when it exists? Describe the stream flow regime (pattern of high and low flows) downstream from your pond, as compared with what it would be if your pond were not built.

8. Add up your pond's total annual evaporation in in./mo, and multiply by the pond's area to get annual Et volume in ac-in./mo. Evaporation is often referred to as a "consumptive" use of water, because water that is lost to the atmosphere is no longer available for direct human use on the site. If your extended-detention pond is built, who would benefit from this consumptive use of water? Find out the principles of water-rights law in your area. Under what conditions would a developer be permitted to install this pond and initiate its consumptive use?

9. Considering all the possible effects of stormwater and what design can do with it, is a bigger volume of runoff following development necessarily good or necessarily bad? What types of site-specific factors could make a large stormwater volume an advantage or a disadvantage?

10. What site conditions are likely to favor infiltration over extended detention as a means of water quality control? What could happen if you disregard these conditions?

SUPPORT OF IRRIGATION

Irrigation is another way to put harvested water to use. "Runoff irrigation" has been implemented notably in Arizona and Colorado, where dry climates cause great demand for irri-

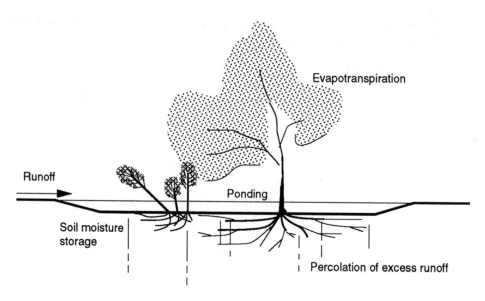

Figure 10.4 Flows and storages in a runoff-irrigated area.

gation water in rapidly growing cities. Irrigated plantings have included lawns, gardens, shrub areas, and fruit and shade trees.

The simplest and cheapest runoff irrigation technique is to direct runoff into planted basins or swales without formal storage or treatment (Figure 10.4). Trees and shrubs rooted in the bottoms and sides of earthen depressions tolerate temporary ponding of runoff while it infiltrates and becomes available to plant roots. To prevent runoff from ponding so long as to be intolerable, the basin should be designed for a limited ponding time as is an infiltration basin. The plants selected for runoff irrigation should be tolerant both of occasional ponding and of dry seasons when runoff is not available.

In wet periods, excess runoff overflows the planted depression and is not made available to plants. As it relates to irrigation, overflowing water is lost. Only the amount of water that infiltrates the soil and stays in the rooting zone without percolation downward is effectively applied to the plants.

Runoff water stored in tanks and cisterns can extend runoff supplies into dry periods between storms. Although the additional hardware increases expense, its use enhances efficiency of use of the available runoff water by reducing overflows and saving water until it is needed. The necessary size of a storage reservoir can be estimated by cumulatively summing its monthly inflows and outflows over a year; the largest amount in storage in any one month is the required volume. Ponds can be used for storage where the edges are designed to tolerate fluctuation in water level from season to season.

Municipal or well water can further supplement the runoff supply. In dry periods, moisture sensors can activate automatic irrigation systems that tide the plants over until runoff comes again.

Collecting and storing water for irrigation affects storm flows the same way infiltration does. If 100 percent of storm runoff is collected, then downstream Q_{vol} and q_p become zero. Some of the water applied for irrigation may reappear in stream base flows after passing slowly through the soil and groundwater.

Several water-harvesting demonstration projects in Arizona are publicly accessible. In Tucson, Casa del Agua is operated by the University of Arizona's Office of Arid Land Studies (Foster, Karpiscak, and Brittain, 1988; Karpiscak, Foster, and Schmidt, 1990). Paved walkways, patios and driveways slope gently to plantings of grapevines and native shrubs. Nearby earthen berms contain the runoff and direct it to the plants. Supplemental irrigation comes, when needed, mainly from residential graywater from sinks, showers, and washing machines. In the planted soil, the graywater's soaps and other mild contaminants are diluted by harvested runoff and treated through biodegradation. Roof runoff is stored in a tank for evaporative cooling and toilet flushing.

In Phoenix, the Arizona Public Service Company's Environmental Showcase Home combines harvested runoff with recycled potable water for irrigation (Pijawka and Shetter, 1995). On the main roof of the house peripheral gutters and hanging chains guide runoff to a buried storage tank, where it joins household graywater. The mixed water irrigates small turf and shrub areas through an automatic irrigation system. On the garage roof, a central gutter flows into an aboveground cistern from which water can be released manually to an adjacent vegetable garden. Excess water flows directly onto adjacent soils for infiltration and supplemental flood irrigation. Also in Phoenix, the Desert House on the grounds of the Desert Botanical Garden (Karpiscak, Brittain, and Foster, 1994) similarly channels roof runoff to adapted plantings and selected interior uses, supplementing graywater and city water supplies.

In Englewood, Colorado, the Greenwood Plaza office park captures runoff for irrigation in several small ponds prominently located at road intersections and building facades (Figure 10.5). Capturing the runoff helps satisfy local stormwater control standards while ensuring that any runoff from excessively applied irrigation water is not lost. Rocky pond edges

Figure 10.5   Greenwood Plaza, Englewood, Colorado.

tolerate fluctuations in water level. Pumps regularly move the water into ponds nearest the areas that need irrigation; at the proper time additional pumps push the water into irrigation pipes for delivery to plants (Horanic, 1980).

SUMMARY AND COMMENTARY

It is paradoxical that contemporary cities shunt away the rain that falls on them while they import fresh water from distant streams and reservoirs.

Conservation and use of the water we are already managing may be the most sustainable way to maintain and extend urban water supplies. By making positive use of on-site rainfall and runoff, water harvesting reduces the need for imported water. It turns the liability of stormwater impact into the asset of water supply.

Using runoff as a water supply requires an extraordinary integration of site layout, contouring, planting, and drainage. In constructing this integration, the results of hydrologic calculations are only parameters for design, not the design itself. As long as you meet the few required quantitative criteria, you are free to design the site any way that is needed in terms of planting, contouring, and materials.

REFERENCES

Arizona Water Resource, 1994, What's in the Rain Barrel? *Arizona Water Resource* vol. 3, no. 5, p. 3.

Athens, Lucia, and Bruce K. Ferguson, 1996, Water Issues, Chapter 6, *Sustainable Building Technical Manual, Green Building Design, Construction, and Operations*, pp. III.13–III.25, Washington: Public Technology.

College of Agriculture, (n. d.), *Water Harvesting Systems*, Tucson: University of Arizona College of Agriculture.

Ferguson, Bruce K., 1987a, Urban Stormwater Harvesting: Applications and Hydraulic Design, *Journal of Environmental Management* vol. 25, pp. 71–79.

Ferguson, Bruce K., 1987b, Water Conservation Methods in Urban Landscape Irrigation: An Exploratory Overview, *Water Resources Bulletin* vol. 23, no. 1.

Ferguson, Bruce K., 1988, Using Water Effectively, Chapter 3 in *Irrigation*, vol. 3 of *Handbook of Landscape Architectural Construction*, Washington: Landscape Architecture Foundation.

Ferguson, Bruce K., 1990a, Assuring Runoff Water Supply to Urban Ponds and Wetlands, in *Proceedings of CONSERV 90*, pp. 281–285, Dublin, Ohio: National Water Well Association.

Ferguson, Bruce K., 1990b, Role of Long-Term Water Balance in Design of Multiple-Purpose Stormwater Basins, in *Proceedings of 1989 Conference of Council of Educators in Landscape Architecture*, Sara Katherine Williams and Robert R. Grist, editors, pp. 161–170, Washington: Landscape Architecture Foundation.

Fink, Dwayne H., and William J. Ehrler, 1984, The Runoff Farming Agronomic System: Applications and Design Concepts, *Hydrology and Water Resources in Arizona and the Southwest* vol. 14, pp. 33–40.

Foster, Kennith E., Martin M. Karpiscak, and Richard G. Brittain, 1988, Casa del Agua: A Residential Water Conservation and Reuse Demonstration Project in Tucson, Arizona, *Water Resources Bulletin* vol. 24, no. 6, pp. 1201–1206.

Frasier, Gary W., and Lloyd E. Myers, 1983, *Handbook of Water Harvesting*, Agriculture Handbook 600, Washington: U.S. Department of Agriculture.

Horanic, Joie, 1980, Re-Using Scarce Waters for Suburban Offices, *Landscape Architecture* vol. 70, no. 4, pp. 389–391.

Jenkins, D., and F. Pearson, 1978, *Feasibility of Rainwater Collection Systems in California*, Contribution No. 173, Berkeley: University of California, California Water Resources Center.

Karpiscak, Martin M., Richard G. Brittan, and Kennith E. Foster, 1994, Desert House: A Demonstration/Experiment in Efficient Domestic Water and Energy Use, *Water Resources Bulletin* vol. 30, no. 2, pp. 329–334.

Karpiscak, Martin M., Kennith E. Foster, and Nancy Schmidt, 1990, Residential Water Conservation: Casa del Agua, *Water Resources Bulletin* vol. 26, no. 6, pp. 939–948.

Pijawka, K. David, and Kim Shetter, 1995, *The Environment Comes Home*, Tucson: University of Arizona Press.

Waller, D. H., 1989, Rain Water—An Alternative Source in Developing and Developed Countries, *Water International* vol. 14, pp. 27–36.

EPILOGUE

The role of landscape architecture and site engineering is in the land, and their obligation is to the land and the people who live with it. Design is responsible for adapting, and can adapt, to the climates, topographies, and inhabitants of specific sites. The goal is to solve human and environmental problems by whatever method is supported by the evidence.

The design skills obtained by studying quantitative hydrology allow you to implement given stormwater programs and objectives. This is essential for full and competent professional practice.

But your skills also allow you to transcend unfounded conventions and to practice directly and originally on every site. By applying your skills vigorously and intelligently, you can interpret what you see happening on a site, discover the hidden facts, evaluate conflicting proposals, marshal evidence, slice through to the systemic sources of problems, overcome contradictory claims, and build new integrations of human communities and natural processes. You need not comply blindly with uniform standards; you need not be a slave to rumor and speculation. You can discover, choose, and advocate.

When I was studying landscape architecture under Ian McHarg and Narendra Juneja in the 1970s, we students took it for granted that we were obligated to understand how each system works before laying our hands on it, and to use the conditions and processes of each site as the basis of design. That principle seems intuitively as right today as it did then. Each site is an organism dependent on balanced throughflows of water, energy, and nutrients.

It is the responsibility of every practitioner to maintain the state of the practice at the state of the science. We cannot sit idly by while the diseases of impervious surfaces and urban sprawl run their destructive courses unchecked and unmodified.

It is time to stop managing stormwater. It is time to start restoring it. The measure of success is the health of the landscape, not the size of the pipes that drain it.

APPENDIX

EXERCISE SITES

The exercises in this book are for you to practice applying mathematical models and design procedures. You can use sites in your region that you are familiar with or that you have come across in your work.

If you prefer, you can select from the sites described in this appendix. Although the sites listed here are hypothetical, they are based on conditions of soil, climate, and topography that actually occur in their locales.

Contrast between two sites that are different in key respects, such as rainfall and soil, stimulates valuable discussion about how specific site conditions create different hydrologic effects and constrain stormwater designs. In my courses, each participant chooses one site from a pair and completes the exercises for that site. Thus, the class as a whole generates two sets of solutions that we can compare and discuss.

The conditions listed in Table A.1 apply to all sites listed in this appendix. The following sections describe unique conditions in regional pairs of sites.

TABLE A.1 Conditions applying to all exercise sites listed in this appendix

Site

Design storm recurrence interval	10 yr
Drainage area A_d	10 ac
Hydraulic length	1,000 ft
Proposed land use	Office or multifamily: 50% impervious + 50% turf
Cover along hydraulic length after development	50% impervious + 50% turf

Conveyance

Slope of swale or pipe	Equal to slope along hydraulic length
Swale material	Grass mixture
Maximum allowable depth of flow in swale	2.0 ft
Maximum allowable width of flow in swale	15 ft
Length of culvert	50 ft
Maximum allowable depth at culvert mouth	2.0 ft

Detention

Detention requirement	q_p after development $\leq q_p$ before development
Depth (head) at maximum flood storage	2.0 ft

Extended detention

Dry-basin extended detention volume	1/2 in. of runoff
Dry-basin drawdown time	24 hr
Wet-basin sediment trap efficiency	66%
Wet-basin inflow	$q\Sigma$, the sum of direct runoff and base flow
Design settling velocity V_s	1 ft/day
Sediment storage volume	Equivalent to 1/2 in. of watershed runoff

Infiltration

Maximum ponding time t_p	1 day (24 hours)
Void space V_d of open-graded stone aggregate	0.40 af/af
Infiltration rate safety factor S_f	0.5
Area B of site available for vegetated basin	10% of site area = 0.10 ac/ac
Area B of site available for subsurface basin	10% of site area = 0.10 ac/ac

Water harvesting

Infiltration rate safety factor S_f	2.0
Pond inflow	$q\Sigma$, the sum of direct runoff and base flow

NORTHEASTERN SITES

The Northeast is familiar with urbanization and the problems it can bring, but many growing locales still need to learn how to urbanize well. On the Coastal Plain, level topography and sometimes sandy soils overlie major aquifers. On the glaciated Appalachian plateau, sedimentary hills are mantled with fine-textured till. Nevertheless, the two areas have more or less similar rainfall. Contrasts in runoff and required drainage facilities are primarily due to differences in soil and slope. Differences in drainage facilities that result from those conditions are sometimes not what one would expect.

TABLE A.2. Northeastern sites

	Plateau Site	Coastal Plain Site
Location	Appalachian Plateau	Atlantic Coast
Weather station	Ithaca, New York	Atlantic City, New Jersey
Physiography	Glacial till	Coastal Plain
Soil series	Bath	Evesboro
Hydrologic soil group	C	A
Slope along hydraulic length	8 %	1 %
Land use before development	Cedar-oak woods	Pine-oak woods

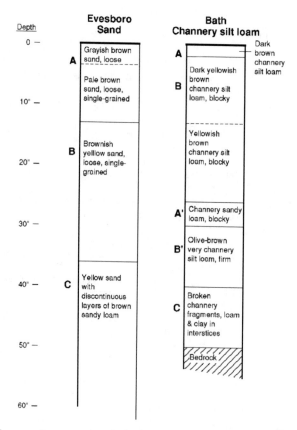

Figure A.1   Soils on northeastern sites (U.S. Soil Conservation Service, 1965, p. 198, and 1978a, p. 16).

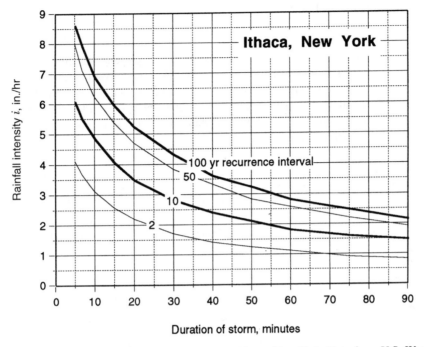

Figure A.2 Intensity-duration-frequency curves for Ithaca, New York (data from U.S. Weather Bureau, 1955).

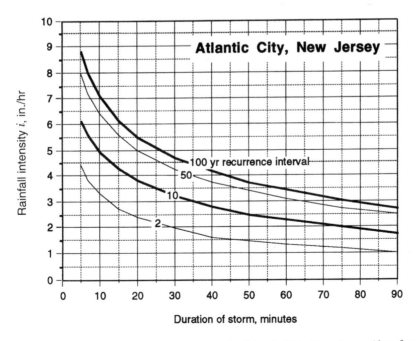

Figure A.3 Intensity-duration-frequency curves for Atlantic City, New Jersey (data from U.S. Weather Bureau, 1955).

SOUTHEASTERN SITES

The Southeast is urbanizing rapidly and is trying to find out how to do it well. The Coastal Plain physiographic region is characterized by level topography and, sometimes, sandy soils overlying significant aquifers. The Piedmont region is characterized by low hills and red clayey soil overlying impermeable crystalline rock. Nevertheless, the two areas have essentially similar rainfall. Contrasts in runoff and required drainage facilities are primarily due to differences in soil and slope. Differences in drainage facilities that result from those conditions are sometimes not what one would expect.

TABLE A.3 Southeastern sites

	Piedmont Site	Coastal Plain Site
Location	Southeastern Piedmont	Atlantic Coast
Weather station	Atlanta, Georgia	Charleston, South Carolina
Physiography	Piedmont	Coastal Plain
Soil series	Cecil	Bonneau
Hydrologic soil group	B	A
Slope along hydraulic length	4 %	1 %
Land use before development	Pine-oak woods	Pine-oak woods

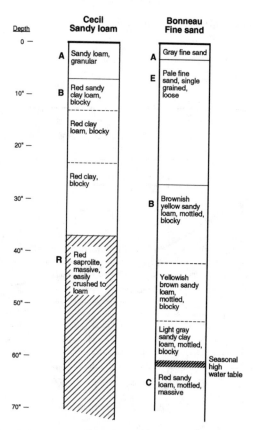

Figure A.4   Soils on southeastern sites (Perkins, 1987, pp. 82–105; U.S. Soil Conservation Service, 1982, pp. 65–66).

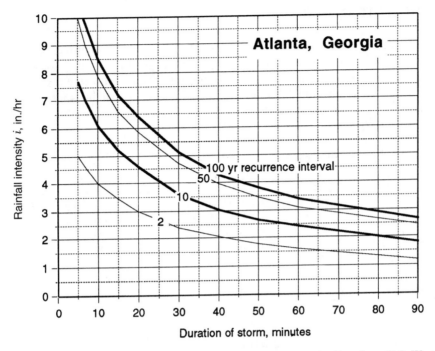

Figure A.5 Intensity-duration-frequency curves for Atlanta, Georgia (data from U.S. Weather Bureau, 1955).

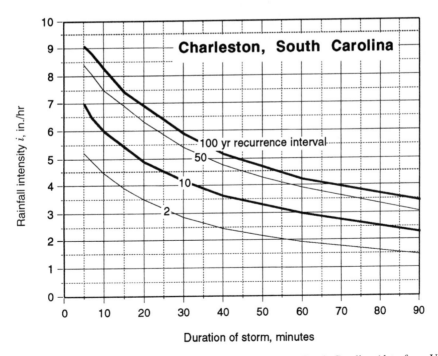

Figure A.6 Intensity-duration-frequency curves for Charleston, South Carolina (data from U.S. Weather Bureau, 1955).

NORTH-CENTRAL SITES

The glaciated plains have been the home of some of America's best-known environmentalists and designers. Here outwash plains have gentle topography and sandy soils with significant groundwater. Till plains have a rolling topography with fine-textured soils. Differences between them in hydrology and required drainage facilities result primarily from differences in soil and slope.

TABLE A.4 North-Central sites

	Outwash Site	Till Site
Location	Lake Mendota area	Lake Michigan area
Weather station	Madison, Wisconsin	Chicago, Illinois
Physiography	Glacial outwash	Glacial till
Soil series	Dickinson	Frankfort
Hydrologic soil group	A	C
Slope along hydraulic length	2 %	5 %
Land use before development	Prairie grass	Prairie grass

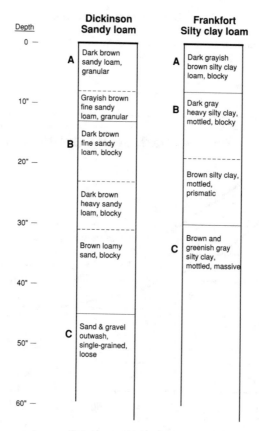

Figure A.7 Soils on north-central sites (U.S. Soil Conservation Service, 1978b, pp. 19–20, and 1979, pp. 95–96).

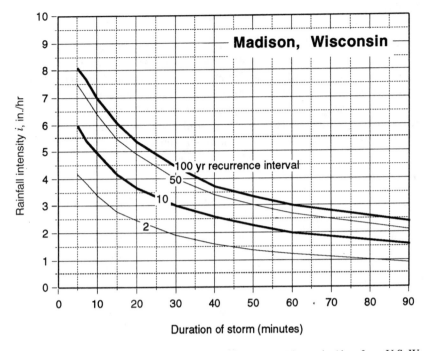

Figure A.8   Intensity-duration-frequency curves for Madison, Wisconsin (data from U.S. Weather Bureau, 1955).

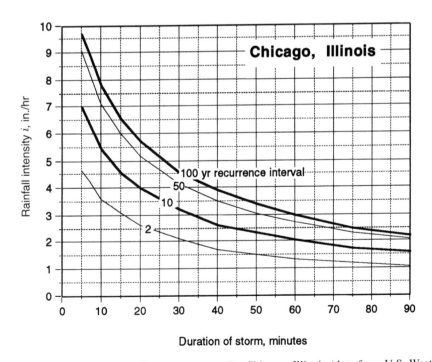

Figure A.9   Intensity-duration-frequency curves for Chicago, Illinois (data from U.S. Weather Bureau, 1955).

TEXAS SITES

The east and west ends of Texas present a dramatic contrast of humid and arid climates. On the Gulf Coast, a constant flow of moisture comes with the southern winds. In the inland west is the desert. However, sites that are more or less similar in slope and soil hydrology can be found in both regions. Among similar sites, differences in stormwater hydrology and management are a result mainly of differences in rainfall and evaporation. Soil characteristics peculiar to these extreme regions, such as shrink-swell potential and high lime content, may require particularly detailed refinements in environmental design.

TABLE A.5 Texas sites

	Eastern Site	Western Site
Location	Gulf Coast	Rio Grande Valley
Weather station	Houston, Texas	El Paso, Texas
Physiography	Coastal plain	Alluvial plain
Soil series	Segno	Hueco
Hydrologic soil group	C	C
Slope along hydraulic length:	1 %	1 %
Land use before development	Pine-oak woods	Desert shrub

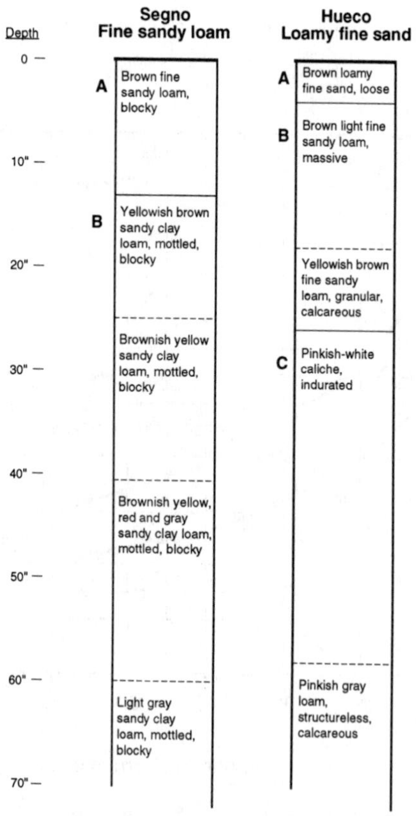

Figure A.10   Soils on Texas sites (U.S. Soil Conservation Service, 1971, p. 24, and 1976b, pp. 56–57).

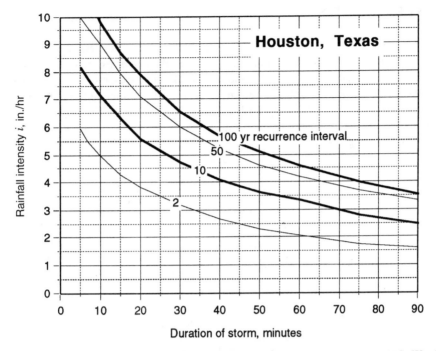

Figure A.11 Intensity-duration-frequency curves for Houston, Texas (data from U.S. Weather Bureau, 1955).

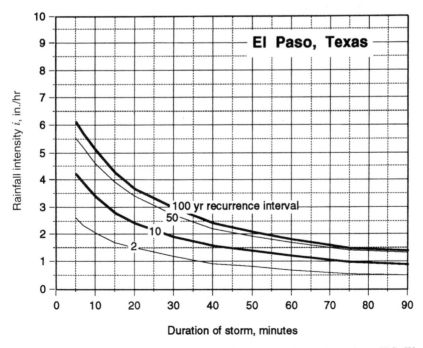

Figure A.12 Intensity-duration-frequency curves for El Paso, Texas (data from U.S. Weather Bureau, 1955).

FRONT RANGE SITES

The Front Range region straddles the boundary between the short-grass prairie and the Rocky Mountains. Development around the growing urban areas extends in all directions, including up the slopes at the feet of the mountains. The benches on the slopes are old alluvial deposits, or are at least capped with alluvial deposits, that emigrated from the mountain canyons. In contrast, the relatively level prairies are often capped with wind-deposited material. The two areas contrast in slope and soil, with correspondingly different results in hydrology and design.

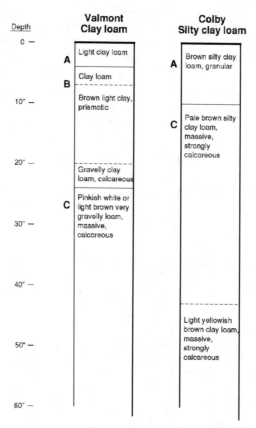

Figure A.13   Soils on Front Range sites (U.S. Soil Conservation Service, 1976a, pp. 9 and 26–27).

TABLE A.6 Front Range sites

	Upland Bench	Prairie Site
Location	Front Range	Front Range
Weather station	Denver, Colorado	Denver, Colorado
Physiography	Alluvial bench	Eolian
Soil series	Valmont	Colby
Hydrologic soil group	C	B
Slope along hydraulic length	6 %	2 %
Land use before development	Prairie grass	Prairie grass

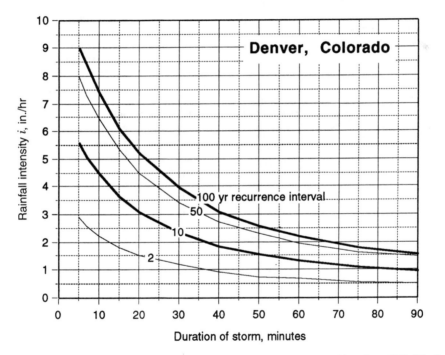

Figure A.14 Intensity-duration-frequency curves for Denver, Colorado (data from U.S. Weather Bureau, 1955).

DESERT SITES

The powerful relief of the Basin and Range province results in sharp physiographic divisions. Sloping alluvial fans drape the feet of the mountains; farther from the mountains the alluvium forms broad plains. The low rainfall of the Southwest requires small drainage facilities, which is convenient in regard to cost. But aridity limits the potential uses that can be made of harvested water.

TABLE A.7 Desert sites

	Alluvial Plain	Alluvial Fan
Location	Rio Grande Valley	Salt River Valley
Weather station	Albuquerque, New Mexico	Phoenix, Arizona
Physiography	Piedmont alluvium	Stream terrace
Soil series	Madurez	Ebon
Hydrologic soil group	B	C
Slope along hydraulic length	2 %	6 %
Land use before development	Pinyon-juniper	Desert shrub

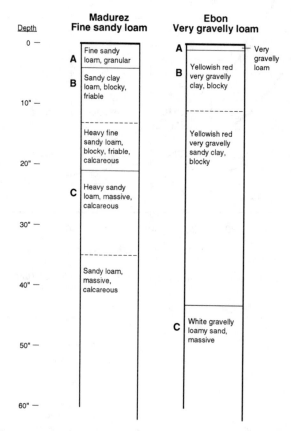

Figure A.15 Soils on desert sites (U.S. Soil Conservation Service, 1977, pp. 26–27, and 1986, p. 125).

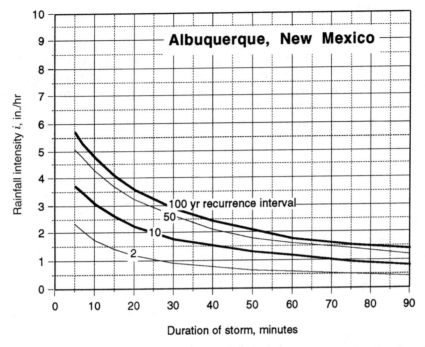

Figure A.16 Intensity-duration-frequency curves for Albuquerque, New Mexico (data from U.S. Weather Bureau, 1955).

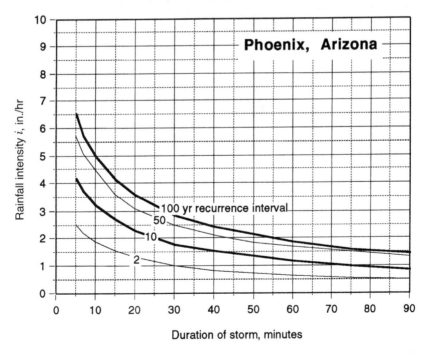

Figure A.17 Intensity-duration-frequency curves for Phoenix, Arizona (data from U.S. Weather Bureau, 1955).

PACIFIC SITES

The Pacific Coast presents a climatic and ecological gradient, with winter rainfall increasing northward. Vegetation increases correspondingly in stature and productivity. Nevertheless, sites with hydrologically more or less similar soils can be found in the alluvial valleys all along the coast, and large storms are surprisingly similar even in zones of different annual rainfall. The ability to take advantage of water as a resource for the environment and for people depends on where a site is located along the climatic gradient.

TABLE A.8 Pacific sites

	Southern Site	Northern Site
Location	San Francisco Bay	Willamette Valley
Weather station	San Francisco, California	Portland, Oregon
Physiography	Alluvial terrace	Alluvial terrace
Soil series	Danville	Powell
Hydrologic soil group	C	C
Slope along hydraulic length	2 %	2 %
Land use before development	Chaparral	Douglas fir woods

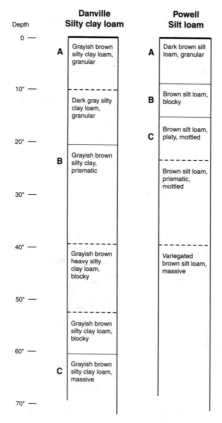

Figure A.18   Soils on Pacific sites (U.S. Soil Conservation Service, 1981, pp. 44–45, and 1987, p. 182).

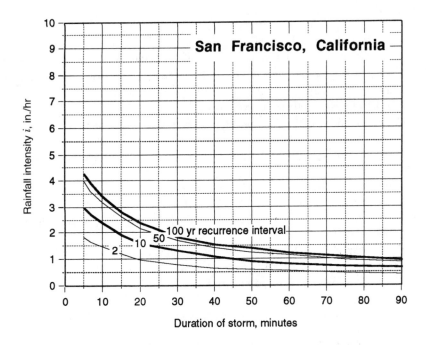

Figure A.19 Intensity-duration-frequency curves for San Francisco, California (data from U.S. Weather Bureau, 1955).

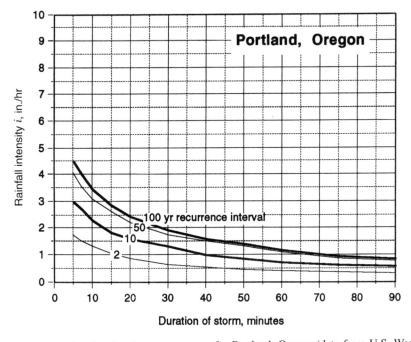

Figure A.20 Intensity-duration-frequency curves for Portland, Oregon (data from U.S. Weather Bureau, 1955).

REFERENCES

Hansen, Wallace R., and Eleanor J. Crosby, 1982, *Environmental Geology of the Front Range Urban Corridor and Vicinity, Colorado*, Professional Paper 1230, Washington: U.S. Geological Survey.

Hunt, Charles B., 1974, *Natural Regions of the United States and Canada*, San Francisco: Freeman.

Perkins, Henry F., 1987, *Characterization Data for Selected Georgia Soils*, Special Publication 43, Athens: Georgia Agricultural Experiment Station.

U.S. Soil Conservation Service, 1965, *Soil Survey, Tompkins County, New York*, Washington: U.S. Soil Conservation Service.

U.S. Soil Conservation Service, 1971, *Soil Survey, El Paso County, Texas*, Washington: U.S. Soil Conservation Service.

U.S. Soil Conservation Service, 1976a, *Soil Survey of Boulder County Area, Colorado*, Washington: U.S. Soil Conservation Service.

U.S. Soil Conservation Service, 1976b, *Soil Survey of Harris County, Texas*, Washington: U.S. Soil Conservation Service.

U.S. Soil Conservation Service, 1977, *Soil Survey of Bernalillo County and Parts of Sandoval and Valencia Counties, New Mexico*, Washington: U.S. Soil Conservation Service.

U.S. Soil Conservation Service, 1978a, *Soil Survey of Atlantic County, New Jersey*, Washington: U.S. Soil Conservation Service.

U.S. Soil Conservation Service, 1978b, *Soil Survey of Dane County, Wisconsin*, Washington: U.S. Soil Conservation Service.

U.S. Soil Conservation Service, 1979, *Soil Survey of Du Page and Part of Cook Counties, Illinois*, Washington: U.S. Soil Conservation Service.

U.S. Soil Conservation Service, 1981, *Soil Survey of Alameda County, California, Western Part*, Washington: U.S. Soil Conservation Service.

U.S. Soil Conservation Service, 1982, *Soil Survey of Colleton County, South Carolina*, Washington: U.S. Soil Conservation Service.

U.S. Soil Conservation Service, 1986, *Soil Survey of Aguila-Carefree Area, Parts of Maricopa and Pinal Counties, Arizona*, Washington: U.S. Soil Conservation Service.

U.S. Soil Conservation Service, 1987, *Soil Survey of Lane County Area, Oregon*, Washington: U.S. Soil Conservation Service.

U.S. Weather Bureau, 1955, *Rainfall Intensity-Duration-Frequency Curves*, Technical Paper No. 25, Washington: U.S. Weather Bureau.

Way, Douglas A., 1978, *Terrain Analysis*, second edition, Stroudsburg. Pa.: Dowden, Hutchinson and Ross.

INDEX